Clinical Genomics

Marcelo A. Marti

Clinical Genomics

Concepts and Applications

Springer

Marcelo A. Marti
Department of Biological Chemistry
School of Exact and Natural Sciences
University of Buenos Aires
Buenos Aires, Buenos Aires, Argentina

ISBN 978-3-032-34666-7 ISBN 978-3-032-34667-4 (eBook)
https://doi.org/10.1007/978-3-032-34667-4

This Springer imprint is published by the registered company Springer Nature Switzerland AG
The registered company address is: Gewerbestrasse 11, 6330 Cham, Switzerland

This book is dedicated to my family, Veronika, Ivan, Joaquin, and Indra, the loudest cheerleaders and the best distractions.

Foreword

The completion of the Human Genome Project marked the beginning of a revolution, but the true challenge—and the true triumph—of the last decade has been the "last mile": bringing that data to the patient's bedside. We find ourselves in an era where the bottleneck is no longer the acquisition of genetic data, but its interpretation and clinical application. In this context, *Clinical Genomics* is not just a textbook; it is a bridge. The field of genomics moves at a breathtaking pace. For the clinician, the researcher, and the student, staying current requires more than just access to journals—it requires a structured framework to organize the vast complexities of NGS, bioinformatics, and variant classification. Marcelo A. Marti has spent years at the forefront of this educational frontier, most notably through the Clinical Genomics School. His ability to distill the intricacies of Mendelian genetics, cancer genomics, and the ethical landscapes of modern medicine into a cohesive narrative is what sets this work apart. What makes this volume particularly vital is its roots in the classroom. This is a text born from the rigorous "trial by fire" of teaching hundreds of students across many editions of a specialized school. It addresses the real-world questions asked by professionals in biochemistry, molecular biology, and medicine, and provides the clarity needed to navigate the standards set by the American College of Medical Genetics and other global bodies. Furthermore, this work serves as a cornerstone for the growing community of genomic professionals in Latin America. By providing a clear roadmap for the integration of genomic insights into clinical care, Marti is not just educating readers; he is helping to build the infrastructure for the future of personalized medicine in the Spanish-speaking world and beyond. For anyone looking to move beyond theory and into the practical, life-saving reality of clinical genomics, this book is an indispensable guide.

Jonathan Zaiat, PhD.
National Institute of Biological Chemistry,
Buenos Aires, Argentina.

Preface

Clinical genomics is an emerging discipline that aims to translate the discoveries and methodologies of genomics into clinical practice, encompassing prevention, diagnosis, and treatment. For readers interested in understanding how the Human Genome Project and next-generation sequencing technologies are reshaping modern medicine, this book offers a comprehensive and clear introduction. Along the journey, we will explore the nuances of DNA sequencing, delve into the dramatic shift that massive sequencing technologies bring to diagnostics, and examine the implications of advancements in bioinformatics. We will investigate the intricate processes of mapping and annotating genetic variations and see how these techniques lead to improved patient outcomes. As we proceed, this book provides deep insights into the application of clinical genomics in the diagnosis and treatment of Mendelian diseases, the impactful classification standards set by the American College of Medical Genetics, and beyond. In examining ethical, legal, and social issues, as well as the promising realms of cancer genomics and population genetics, readers will gain a multifaceted view of how genomic insights are integrated into clinical care to not only treat but also predict and prevent diseases, heralding a new era of personalized medicine.

This book is designed for a broad and diverse readership and is especially well suited for students and professionals in fields such as medicine, biochemistry, molecular biology, biotechnology, and bioinformatics, who are keen to understand the clinical applications of genomics. It also caters to healthcare professionals seeking practical insights into how clinical genomics operates across various medical environments. Furthermore, this volume is an indispensable resource for individuals looking to enhance their knowledge of the role of genomics in disease diagnosis, treatment, and prevention. Educators and academics will find it an excellent addition to curricula that aim to incorporate human genomics, providing a rich trove of data and insights. Additionally, patients interested in understanding how advancements in genomics may affect their treatment options will appreciate the clear and accessible explanations provided. Overall, this book is a comprehensive guide for anyone eager to understand how genomics is reshaping the landscape of modern medicine.

Buenos Aires, Argentina — Marcelo A. Marti

A Personal Note to My Students

I would also like to dedicate this book to the students who have participated in the many (more than 11eleven) editions of the Clinical Genomics School. Within these pages, they will find expanded content covering each topic addressed in both the introductory and advanced programs. It is my hope that this volume serves as a complement to the theoretical and practical training provided in the courses. If, on the other hand, you are a reader who has not yet attended our school and are a Spanish speaker, do not miss the next opportunity by keeping an eye on http://bioinfo.qb.fcen.uba.ar/

Best regards,
Marcelo A. Marti
Professor, Biological Chemistry Department
Facultad de Ciencias Exactas y Naturales de la
Universidad de Buenos Aires
Buenos Aires, Argentina

Acknowledgments

Writing this book has been a journey shared with many. First and foremost, my deepest gratitude goes to the Clinical Genomics School community: Jonathan, Franco, Laura, Anita, Sofi, and many others. To the students of our many editions, your curiosity, challenging questions, and passion for precision medicine were the primary catalysts for transforming our course materials into these chapters. I also want to extend a special thanks to my colleagues and peers who have been instrumental in my growth. Their insights and rigorous discussions have sharpened the concepts presented in this volume. A particular mention must go to the LatAm Clinical Genomics community. To my fellow Spanish-speaking professionals across Latin America, thank you for your commitment to advancing genomic medicine in our region. Our shared goal of bridging the gap between complex data and patient care is what makes this work meaningful. It is an honor to be part of such a vibrant and dedicated network of experts.

I also want to thank the Facultad de Ciencias Exactas y Naturales de la Universidad de Buenos Aires for its constant financial support.

Contents

Epilogue

As we reach the culmination of this exploration into the dynamic realm of clinical genomics, it is imperative to take a moment to reflect on the journey we have embarked upon—from deciphering the fundamental building blocks of human life to the sophisticated application of genomics in clinical settings. The chapters in this book offer not merely a collection of scientific advancements and methodologies but narrate a profound story of discovery, innovation, and the relentless pursuit of knowledge that propels humanity forward. The narrative began with an introduction to clinical genomics, shedding light on its implications for personalized and precision medicine. Moving through the intricacies of DNA sequencing techniques and the transformative impact of bioinformatics, we delved deep into the genetic undercurrents that sculpt our understanding of human health and disease. Alongside the scientific and technological narratives, there also ran a potent thread concerning the ethical, legal, and social implications that accompany the knowledge and capabilities genomic science offers. In the core chapters focused on the application of clinical genomics, we examined its pivotal role in diagnosing and understanding rare Mendelian diseases, explained and discussed the ACMG criteria, and learned how to select and report clinically relevant variants. The latter chapters open dialogues about the challenges and future of fields like cancer genomics and population genomics, emphasizing not only the immense potential but also the profound responsibility borne by researchers, clinicians, and ethicists in leveraging this knowledge responsibly.

As we gaze toward the horizon, the future of clinical genomics sparkles with immense potential, though it is interlaced with intricate challenges. The rapid pace of technological advancements opens the floodgates to a realm of new discoveries and applications that, today, might seem like the stuff of science fiction. Innovations such as gene editing technologies, particularly CRISPR and related systems, are refining our ability to intervene at the very essence of life, offering hope for the cure of genetic disorders that were once deemed untreatable. The integration of artificial intelligence (AI) and machine learning (ML) in clinical genomics is another transformative factor that promises to enhance data analysis, improve diagnostic accuracy, optimize therapeutic interventions, and even predict disease susceptibility with unprecedented precision. These AI systems, for example, are beginning to predict the effects of genetic mutations by analyzing vast datasets of genetic information, which can accelerate the identification of targets for novel drug therapies. Their potential to analyze and interpret the complex patterns of human genomics could revolutionize personalized medicine, making it more accessible and effective. Furthermore, the advent of portable genomic sequencers and AI-enhanced diagnostic tools signifies that high-resolution genomic medicine could become commonplace in routine healthcare, extending even into remote areas where traditional healthcare resources are scarce.

However, this bright vista is not without its shadows. The promise of clinical genomics is inextricably linked with the need for equitable access to these revolutionary treatments and technologies. The disparities in healthcare between different regions and socioeconomic groups could be exacerbated if these technologies are not implemented thoughtfully. Additionally, as our capabilities to manipulate and understand our genetic makeup grow, so too does the imperative to safeguard genetic data against misuse. The protection of such sensitive information and the ethical management of genetic technologies must be upheld to prevent potential discrimination or breaches of privacy. Continuous ethical scrutiny is essential to navigate these waters successfully. International collaborations and frameworks might be needed to handle these global challenges effectively. Only through vigilant regulation, inclusive policies, and ongoing dialogue among scientists, ethicists, policymakers, and the public can we ensure that the advancements in clinical genomics contribute positively to global health without compromising our moral and ethical standards.

In synthesizing the vast information and discussions presented, this book aims to not only inform but also inspire. Whether readers are budding scientists, experienced practitioners, or simply individuals curious about the genetic threads that influence our lives, the journey through clinical genomics is one of profound connectivity and relevance to us all. Thus, as we close this book, let us not consider it an end but a milestone in a continuing journey of discovery and application. Let this be a lit beacon that guides curious minds towards the awe-inspiring realm of genomics, armed with knowledge, ethics, and the unyielding quest for precision in the art and science of healing.

Chapter 1
Introduction to Clinical Genomics

Abstract This chapter introduces Clinical Genomics as a translational discipline that bridges genomic science and medical practice, enabling the application of genomic knowledge to disease prevention, diagnosis, and treatment. It traces the conceptual foundations of genomics, from the discovery of DNA structure to the recognition of the genome as an information system encoded in nucleotide sequences. The chapter explains how sequencing technologies have transformed our ability to read and interpret this information, ushering in the genomic era. Central to Clinical Genomics is the complex relationship between genotype and phenotype, shaped by genetic variants, environmental context, and their interactions. This relationship is explored through key concepts such as effect size, odds ratios, and variant frequency, highlighting differences between rare, high-impact variants and common, low-effect variants. The chapter also outlines major clinical applications, including rare disease diagnosis, oncology, pharmacogenomics, and preventive medicine. Furthermore, it presents the operational framework required to implement clinical genomics in healthcare systems, emphasizing multidisciplinary collaboration, data interpretation, and ethical considerations. Finally, it discusses future directions, including advances in population genomics, artificial intelligence, and gene editing, positioning Clinical Genomics as a central component of modern and future medicine.

Keywords Clinical genomics · Genotype-phenotype relationship · Genetic variants · Sequencing technologies · Precision medicine

What is Clinical Genomics?

Defining a discipline is invariably a challenging endeavor, frequently sparking intense discussions, particularly at the boundaries where it overlaps and interacts with other fields. Moreover, new disciplines do not arise in isolation; they must carve out a

M. A. Marti, *Clinical Genomics*,
https://doi.org/10.1007/978-3-032-34667-4_1

distinct identity from their foundational areas. Clinical Genomics can be defined as an "emerging" discipline dedicated to translating genomic discoveries and methodologies into clinical practice—covering prevention, diagnosis, and treatment. It epitomizes a translational discipline, aiming to move breakthroughs from the laboratory bench directly to the patient's bedside, thereby bridging basic science with its practical medical applications.

Clinical genomics is rooted in the broader field of genomics, especially human genomics, which gained widespread attention in the early twenty-first century following the completion of the Human Genome Project and the advent of Massive Sequencing Technologies (MSTs). We will explore these technologies more thoroughly in subsequent chapters. Consequently, it maintains a tight conceptual and technical connection with genetics and molecular biology. As shown in Fig. 1.1 Clinical genomics comprises five primary branches and unfolds through sequential stages that collectively forge a pathway for its full integration into societal healthcare systems. These stages are: (1) Deciphering the Genome, (2) Understanding the Genome, (3) Linking the Genome with Diseases, (4) Improving Prevention, Diagnosis, and Treatment, and (5) Enhancing the Effectiveness of Healthcare Systems. The journey begins with "pure" genomics; let us delve into what this intriguing starting point entails.

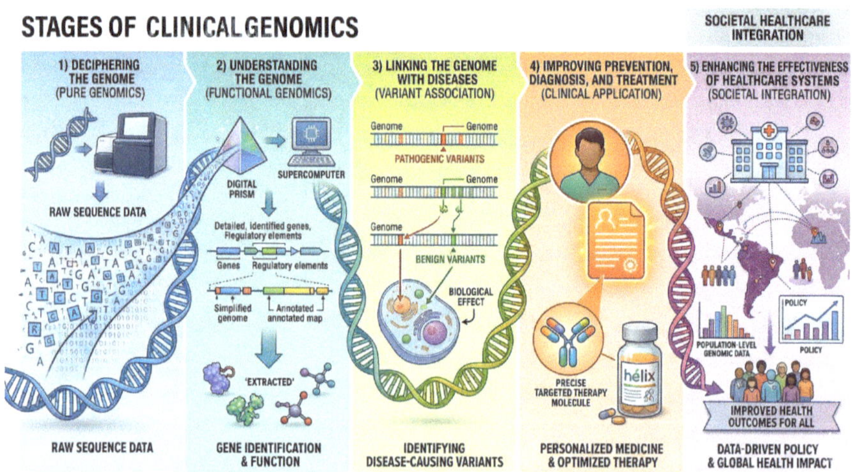

Fig. 1.1 The Five Stages of Clinical Genomics Integration. The pathway toward full integration into healthcare systems involves five sequential phases: (1) Deciphering the Genome, focused on raw sequencing technology; (2) Understanding the Genome, involving functional annotation and gene identification; (3) Linking the Genome with Diseases, the identification of pathogenic vs. benign variants; (4) Improving Prevention, Diagnosis, and Treatment, the direct clinical application through personalized medicine; and (5) Enhancing Healthcare Systems, the final integration into societal policy and global health infrastructure

Genomics

Genomics is the science that studies genomes. A genome, from a conceptual and ontological standpoint, represents the entirety of an organism's genetic information encoded within its DNA, with "information" being the pivotal concept. Let us go back to 1953 when Watson and Crick in their famous work in the journal Nature presented the atomic and molecular structure of DNA. That "double helix," which today we all recognize, has become ubiquitous as an icon in countless biological contexts. In the middle of the 20th century, the scientific community was seeking the definitive answer to the following question: How is genetic information stored and copied from generation to generation? The discovery of the DNA structure proposed an answer, as stated by Watson and Crick in the concluding remark of their work, where the following sentence can be read: "It has not escaped our notice that the specific pairing we have postulated immediately suggests a possible copying mechanism for the genetic material". The structure of the DNA double helix, showing the specific pairing of the bases A:T and C:G, laid the foundations for the discoveries associated with semiconservative DNA replication, and the developments related to the manipulation of recombinant DNA, whose heart is the polymerase chain reaction (PCR). Interestingly, the DNA structure also reveals another fundamental fact of biology, not mentioned in the cited article, which will lead to intense studies for the next 50 years. This "other fact" is the realization that the DNA structure has a high information content. The information that defines a genome.

In the structure of DNA, information is contained in the order of the bases (letters) along a strand, resulting in a specific base sequence (the DNA sequence). To understand how the DNA sequence contains information, let us look at other sequences that also contain information. For example, observe (or read) the following sequence of characters:

"Today is January the 1st, 2024,"

What information does this sequence possess? What type of characters does it contain? How did you interpret or decode it? You will have easily realized that the information provided by the sequence of characters in the example (letters, spaces, numbers), is that today (perhaps the day it was generated, or written) was the 1st day of January of the year 2024. That sequence possesses alphanumeric characters and is written (coded) in English, and you (the reader), if you know English, surely can understand it. Now let's slightly modify the sequence to the following:

"Hoy es 2 de enero de 2024,"

What has changed? Firstly, the information content has changed (can you guess or understand how?). Secondly, the knowledge needed to decode it, has changed. Previously it was English, now it's Spanish. As for the characters, we see that although the ones used are not exactly the same, they still belong to the Western alphabet and numeral system. Let's now modify the first sequence in a different way:

"Today is 01 of January of 11111101000"

It seems nonsensical, doesn't it? To decode it, we must know that now for the numbers conveying the information of the day and year (previously 1 and 2024) the binary

numeral system was used. Asking Google (or ChatGPT) to "translate" 11111101000, will show us that the binary number corresponds to 2024 in decimal notation. This information allows us to know that the content of information, in this case, has not changed (with respect to the first sentence).
Let's now analyze the following sequence of characters:
"AATCGACTTAGTA"
What information does it contain? Can you decode it? The sequence, as you may have guessed, corresponds to a DNA sequence. As such, it is written in the DNA alphabet, which only has four possible letters A, C, T, and G. To decode it, we must know and understand the language of biology (i.e., that of DNA) and be able to translate it into concepts that we can understand.

Understanding and deciphering the language of DNA has been one of the main objectives of the biological sciences for the last seven decades. Today, we know a lot about it. Starting with the genetic code, which allows for the translation of coding sequences into the respective amino acids that form proteins, we also know how to decode where genes start and end, what their internal structure is like (exons and introns), how mRNA is processed, and increasingly also, how gene expression is controlled and regulated, as well as how the supramolecular structure of DNA is organized, among several other things. All the information to control and execute these processes is written in the DNA. Therefore, alongside the process of understanding the language, several techniques have been developed to determine, given a molecule of DNA, the order of its bases, that is, its sequence. The techniques (or methods) used for this purpose are therefore referred to as sequencing methods (or sequencing technologies).

Today, sequencing technologies have acquired, as we will see in Chap. 2, a tremendous capacity to read DNA, and have dramatically reduced the costs and times of the entire process. These facts are the ones that opened the doors to the genomic era. In the coming chapters, we will see what these technological changes are, and what new opportunities they offer. But first, to continue our understanding of Clinical Genomics, let's briefly address the third aspect of this discipline that seeks to relate the genome to human (or simply our) health.

The Human Health and Genome Relationship

To begin to address the third aspect of clinical genomics, which seeks to link the genome with our health, we will start by precisely introducing, or defining, a series of concepts that we will use throughout the book. Let's return to the concept of the genome and place it in the context of an individual person. As we will see in Chap. 3, when we talk about "The" Human Genome, we refer to the typical genome of our species. This genome, which characterizes us as members of the species *Homo sapiens sapiens*, serves as the reference DNA sequence against which we define the genomes of individuals. Each individual person actually possesses a unique genome, which differs in many places from the reference human genome. These differences are technically known as variants. We can therefore define an individual's genotype as the collection of their genetic variants. Note then that the genotype is equivalent to the individual genome, and therefore to the set of an individual's genetic information. If we now define the phenotype, as the set of observed characteristics of the individual (hair color, height, lactose intolerance, AB positive blood type, aerobic capacity, IQ level, and everything that can be observed and/or measured), we will see that what we seek to discover is, the relationship between the variants and the aspects of the phenotype associated with, or related, to our health. In other words, the health-related genotype-to-phenotype relationships.

How is the Genotype Related to the Phenotype?

The relationship between genotype and phenotype is highly complex and not direct. To understand it, we will make use of the widely used resource of analogies, particularly the one proposed by the famous biologist and prolific author Richard Dawkins, who compares the genome to a cooking recipe. In this analogy, a genome is a recipe that contains the instructions for making (in a biological sense) an individual of the species to which that genome belongs. In other words, just as a recipe for making a lemon cake, when followed, allows one to make a "lemon cake," the human genome contains the necessary information to make a human being, the whale genome has the information to make a whale, the eucalyptus genome possesses what is necessary to make a eucalyptus, and the COVID genome holds the information to make more COVID viruses. Finally, it is also important to mention that just as the recipe needs a cook, ingredients, and tools (flour, eggs, oven, mold, etc.) to "make" the cake, the genome requires a suitable biological environment or setting for the development of an individual. In the case of humans, and most mammals, this environment comprises the maternal uterus, in eucalyptus, the seed in fertile soil; and in COVID, a cell of the individual it infects. The proper interaction between the genome (recipe) and environment (cook, ingredients, etc.) is crucial for the healthy development and maintenance of an individual. Progression toward disease usually occurs when there is a "failure" in either the genome or its environment. Continuing with our analogy, we can envision how a mistake in the recipe will produce a defective cake. For example, if when copying a recipe from the internet the word sugar is accidentally changed to the word salt, the cake will likely be unpalatable. Or, if the instruction to add baking powder is omitted, it will have an inappropriate consistency. Similarly, if the genome is significantly altered or an important instruction is omitted, the result will be an individual considered sick or unhealthy. It is also clear, that even with the correct instructions (recipe or genome), a wrong interaction with the environment will produce a defective result. For example, too much cooking time will yield a burnt cake, and excessive exposure to sunlight will possibly cause skin cancer. A second level of complexity in the genotype-phenotype relationship is to observe that the phenotype is not only the result of a specific characteristic of the genotype (an ingredient), but of the interaction between different aspects of the genome (or recipe). And continuing with our analogy, just as we cannot associate the texture of the cake to a particular ingredient or instruction, often a particular pathology (or symptom) cannot be associated with a specific variant.

How Do We Characterize and Quantify the Relationship Between Genotype and Phenotype?

Quantifying the relationship between genotype and phenotype involves measuring the contribution of a given genotype variant to a given phenotype or trait. In this quantification, it is assumed that, except for the studied variant, the rest of the genotype does not have a significant contribution to the trait of interest. Additionally, we will assume that all other characters of the phenotype are not significantly affected by the presence (or absence) of the variant. In other words, the genotype-phenotype relationship becomes a variant-trait relationship, and they are quantified independently from other variant-trait relationships. To quantify (mathematically) this relationship, what is measured is the differential effect observed due to the presence of the variant, on the trait of interest. This magnitude is often expressed as an Odds Ratio (OR) when assessing the association between a variant and a binary trait. Although OR is not the only possible measure of effect size, it is the most commonly used in genetic association studies (which we will explain in detail in later chapters). This magnitude is a value that should be interpreted as the probability (or likelihood) of possessing the trait, given that the variant is present in the individual. This probability (the Odds Ratio) is actually a probability relative to the population average, which means it tells us how many times more (or less) likely, we are to possess the trait (if we have the variant), when compared to an average individual in the population. This magnitude can take only positive values, ranging from zero to infinity. Let's look at some characteristic numbers.

Case 1: An Odds Ratio value of 1 indicates that one has the same chances as an individual from the general population, which is equivalent to saying that the variant has NO effect on the studied trait.

Case 2: An Odds Ratio value greater than 1 indicates that one has an increased probability of possessing the trait, given the presence of the variant. For example, for a variant associated with Parkinson's disease with a value of 2, it is interpreted as the variant doubling the chances of developing Parkinson's disease. Similarly, a variant with an Odds Ratio value of less than 1, indicates a lower probability of having the trait when possessing the variant.

Case 3: An Odds Ratio value tending to infinity means that the presence of the variant is determinant in relation to the trait development. In other words, having the variants results in always developing or possessing the trait.

Let's then see what values we can expect according to the different traits associated with human health. Very high values (tending to infinity) are observed for variants related to classical genetic diseases. Their infinite effect size results in saying the variants are the cause of the pathology. Variants with intermediate values, (ranging from 2 to 10), are observed in pathologies with a high genetic contribution, but where the environment and the interaction between loci play an important role. In this range, we can find variants associated with neurodegenerative diseases (Alzheimer's) and some familial forms of cancer, among others. Finally, low values, 0.5–2, are associated with pathologies called multifactorial, where the environment and interactions between loci are fundamental, such as cardiovascular or metabolic diseases.

The different ranges of values in the effect size are the consequence of different underlying biological mechanisms in the link between the variant and the trait (i.e., different underlying biological genotype to phenotype relationships), and therefore require different treatment in their implementation in clinical genomics, as we will develop and explain specifically in later chapters.

Variant Frequency

As previously mentioned, each individual has his/her own unique genotype, that is, his/her own collection of variants. However, variants are not exclusive to the individual and can also be present in other people. This may be either because both individuals have inherited the same variant from a common ancestor or, because the variant has independently appeared in both lineages (or individuals) purely by chance. When we measure the percentage of the human population that possesses a given variant, this results in the frequency of the variant in the studied population. The frequency of a variant is one of its most important characteristics, and at this point, we want to relate it to the previously mentioned effect size of the variant for the types of pathologies described previously. Generally, variants with large deleterious effects tend to be rare due to negative selection, whereas variants with small effects can drift to higher frequencies. Thus, variants with infinite effects usually have extremely low frequencies. They are thus very "rare" (infrequent), and this explains why they tend to be the cause of Rare Diseases. At the other end, high-frequency variants, with values up to 50%, usually have very mild effects, and are associated with the aforementioned multifactorial diseases (cardiovascular, metabolic, sporadic cancers and neurological diseases).

With this information, we can conclude the genotype-phenotype relationship section clarifying some terms that are used in the literature, especially in non-specific contexts, as synonyms for variants. We should try to avoid them, due to their lack of precision, unless in descriptive cases and when the conditions that define them according to customary uses are clearly met. These are: Mutation, which is usually used to refer to a low-frequency variant with a significant (tending to infinite) and deleterious effect on the phenotype. Polymorphism: usually used to refer to a relatively high-frequency variant (more than 5%) without a significant phenotypic effect. Finally, the term allele, which is usually used to refer to the reference or alternative genotype (that is, the presence of a variant) for a specific locus. In other words, the reference genotype and the variant are two alleles for the same locus (position in the genome).

Clinical Genomics Applications

Genomics holds enormous potential for integration into many areas of medicine, particularly in diagnosis, treatment, and prevention. Over the past decade, genomics has made major advances in two key medical domains: rare Mendelian diseases and cancer. Although these conditions differ fundamentally in their biological origins, both rely on identifying specific genomic variants that drive the disease. In Mendelian disorders these variants are germline and inherited, whereas in cancer they are somatic and arise during tumor evolution. A genomic diagnosis of a rare genetic disease typically involves sequencing the whole genome or a targeted region to identify the variant(s) responsible for the patient's phenotype. This molecular diagnosis—central to clinical genomics—often provides direct therapeutic implications, enabling variant-specific treatments. In oncology, genomic sequencing plays a parallel role: analyzing the tumor genome allows detection of somatic driver mutations that guide prognosis and therapeutic decisions. In many cancers, identifying these alterations is essential for selecting targeted therapies, much like identifying pathogenic variants is essential for diagnosing Mendelian disorders. Despite their biological differences, both applications illustrate the same principle: clinical genomics uses sequence information to reveal the variants that explain disease and guide individualized treatment.

A second branch of application of clinical genomics in the context of treating several pathologies, is what is known as pharmacogenomics (which we will also discuss in a later dedicated chapter). Pharmacogenomics studies the relationship between variants and the phenotypic characteristics associated with the response to specific pharmacological treatments. In this context, pharmacogenomic studies aim to guide pharmacological treatment based on the observed variants in the patient.

In relation to prevention, there are two quite different branches where genomics makes its impact. On the one hand, in the case of genetic diseases, determining the responsible variant and how it segregates in a family, allows for adequate family planning. Meanwhile, in the diagnosed individual, this knowledge can help prevent the development of specific future symptoms. On the other hand, as we have already mentioned, there is a broad number of variants with Odds Ratios between approximately 0.5–10 for various complex diseases, ranging from those with a greater genetic contribution like breast cancer, to those with a high environmental component like cardiovascular diseases. In these cases (and as we will see later in a specific dedicated chapter), knowing how an individual's genotype predisposes him/her to develop some of these pathologies, allows for preventive or monitoring measures that can prevent, delay, or effectively treat the potential pathology.

Another interesting point of clinical genomics and our health, relates to the following fact. While our genome is one and accompanies us unchanged (with some exceptions) throughout our life, clinical genomics will focus on different aspects of it at different stages of our life. For example, family planning and assessment of the potential development of a child with a genetic disease can be analyzed preconceptionally. During pregnancy, there are an increasing number of "Non-Invasive

Prenatal Testing (NIPT)" genomic tests that allow determining the presence of pathologies of genomic origin, such as those caused by chromosomal alterations, in the developing fetus. From there, we move to the potential neonatal screening of genetic diseases, and in the case of the appearance of symptoms in childhood, to the possible use of clinical genomics for molecular diagnosis. The search for a diagnosis may continue into adulthood, although it is at this stage, and in healthy individuals where focus will begin to shift towards prevention, assessing the risk of the potential development of multifactorial diseases. Finally, sooner or later, everyone will face some pathology (of greater or lesser complexity) and it is there where pharmacogenomics will make its greatest contribution. The relevance of these observations lies in highlighting the importance of our genome as a tool for the care of our health throughout our life, a characteristic that few, or possibly no other health studies can maintain.

Finally, a last comment regarding the applications of clinical genomics relates to the medical specialties where it has impacted, or is believed to do so in the near future. Clearly, as already mentioned, genomics has revolutionized medical genetics, and has enormous and growing relevance in oncology. But currently, in relation to diagnosis-treatment, more and more branches of medicine where a significant number of patients have genetic diseases have been incorporating clinical genomics into their practice. These areas include, among others, endocrinology, immunology, rheumatology, ophthalmology, neurology, and even pediatrics. If we add to this, all the specialties where a strong impact of clinical genomics is expected in relation to the prevention of complex and multifactorial diseases (cardiovascular, metabolic, neurodegenerative), it is possible that this trend will continue its course, and sooner rather than later, the broad majority of medical specialties will have incorporated elements of clinical genomics.

Challenges for the Successful Implementation of Clinical Genomics

The final goal related to the successful development of Clinical Genomics is to improve health systems. This item aims to address various aspects related to the implementation of clinical genomics so that its integration into medical practice, and for society in general, is successful and efficient. This is an area of intense debate that, beyond the scientific issues mentioned earlier involves Political, Economic, Legal, Social, Educational, and Media aspects, among others. While each requires extensive development, and depends greatly on the context (i.e., where and when), in this section, I propose a roadmap that shows the most important variables to consider when seeking to implement clinical genomics.

To organize the explanation, let's first look at a general implementation scheme for Clinical Genomics. The entry point, as in all aspects of medicine, is always the patient. Regarding patients, we must then define what type of patients we are looking to help with clinical genomics. Are they healthy patients? Or do they already have a clinical diagnosis? If they have a diagnosis, what type of pathology do they present (Rare Diseases, Cancer, Cardiovascular)? What age range do the patients belong to—are they neonates, pediatric, or adults? As a counterpart to knowing the type of patient, we must establish what the goal for them will be, within the clinical genomics service. Am I looking to provide a genetic or molecular diagnosis? Am I looking to determine possible pharmacogenomic interactions? Or am I looking to evaluate the genetic risk of developing some pathology in the future? Having defined which of these cases corresponds to the case of interest, this will allow defining which application of the clinical genomics repertoire as described earlier must be developed, and then focus on the associated technical aspects.

The second step involves determining what and how the patient's genome (or part of it) will be sequenced. This step requires understanding what technical options are available, their accessibility (do I have a sequencer available, or must I request an external sequencing service, local or international?) and also their cost and diagnostic efficiency. As we will see later, in Clinical Genomics there is no standard methodology, and there is always a range of possibilities. Generally speaking, there is a direct relationship between the cost and the amount and quality of genomic information obtained. The more we pay, the more genomic information we receive. However, more information does not necessarily imply a better result (in relation to our goals), and therefore, it is necessary to find the point where diagnostic efficiency is maximized. It will comprise that strategy that allows diagnosing the largest number of cases (or patients) at the lowest possible cost per case.

The third step, which we haven't discussed yet but will see in-depth later, includes the processing of genomic data. It suffices at this point, to remember that the result of a sequencing study is a string of letters "A, C, T, Gs" in biological language, and therefore to understand and interpret it we must translate it into English. This translation involves first determining the variants of the genotype under study, and then associating them with the patient's health in the context of the study's objective. Currently, there is also a range of possibilities in relation to how broad (what types of links we want to obtain) and deep (how much I want to know about what is known about the pathological link of each variant). For this process, there is also a direct relationship between cost and amount of information we will receive, and there will be an optimal point of information necessary to achieve the desired result, which allows maximizing diagnostic efficiency. In this step, as in the previous one, market considerations in terms of cost and accessibility are also added, and both must be considered for a successful implementation.

Here we reach the point that is the heart of the implementation of a clinical genomics service, which is: the interpretation of the genotype concerning the observed phenotype and the study objective (i.e., the patient). This step consists, for example in a patient where the goal is a genomic diagnosis of a clinically suspected underlying rare disease, in the determination of the pathology causal variant. Or in a patient (or case) where the goal is to assess the genetic risk of developing type 2 diabetes, to determine the patient's relative risk due to the genetic variants present in his/her genotype, and then interpret it in the context of other variants associated with metabolism. Conducting this interpretation requires deep knowledge of the pathology, its goals, and clinical genomics. This is the main reason why clinical genomics interpretation is performed by multidisciplinary teams, which integrate clinical presentation, molecular findings, variant classification guidelines, and bioinformatic analysis and that usually include Physicians, molecular biologists/biochemists trained in clinical genomics, and bio-informaticians trained in processing and analyzing genomic data. The central aim of this book is to provide these professionals, and others who are interested, with a solid and profound training to perform this analysis.

Finally, the last step of the process consists of preparing the report and providing feedback to the patient (and/or requesting professional). This is a key step as it represents the result of all previous steps and is a reflection of our work. A good report will adequately reflect the quality of the clinical genomics service, and that is why we will also devote time to analyzing what elements, and how it should be prepared.

Having thus understood the process that we must perform to implement a clinical genomics service, we will briefly analyze what contextual variables must be considered for its successful development. We refer to the implementation context, to those variables that define the institutional context of implementation including: which individuals will perform it, which institution, and in which state of country? Examples of these contexts, which show the enormous range of possibilities, are: A Medical Professional in their own office, a clinic, a hospital, a research laboratory at a university, a diagnostic services laboratory, a provincial or national network of the public health system, or even a ministry of health. These contexts define, in a specific way (or within a certain range), several of the variables described in the previous point, such as: the target audience of patients, access to sequencing services and data processing/analysis, and/or the existence of teams for interpretation. Defining these contexts also adds other relevant variables such as: economic-financial aspects (who pays for the service, how much and how), legal-regulatory aspects (how much, how, and where I must store the data, do I have restrictions or not for hiring services to third parties, etc.), Educational (the purpose of the service includes educational aspects for professionals and/or the community), and Media (how the existence of the service and its results are communicated).

In addition to the aspects mentioned per se, each and every one of them must be implemented following the general ethical and legal norms of medical practice, and those specific to genomics (if there are any). In an ideal world, one should have resolved all the aspects mentioned before implementing a clinical genomics service, and that is why during the development of the coursebook we will address each of them in detail. However, they say that "the path is made by walking," and that is why to provide a clinical genomics service it is possible to start with something limited that resolves (or defines) some of the mentioned variables, and then allows expanding the range of possibilities of the service. Our "strong" recommendation in this regard is that one should start with developing the genomic interpretation service, associated with a well-defined group of patients and with a clear and limited goal. The rest of the steps can be outsourced at first, and then be incorporated into the service depending on the possibilities and advancement of the same.

At this point in the reading, we now have all the tools that allow us to understand: What is Clinical Genomics?, What are its potential impacts?, and What do we need to develop it?. But before moving on to the technical aspects related to sequencing technologies (Chap. 2), we want to leave you with two reflections associated with genomics that we believe may be of interest.

Personalized Medicine, Precision Medicine, and Clinical Genomics

For some readers, the term "Clinical Genomics" might not seem the most fitting or impactful to describe the contents of this book, and perhaps it would be better to discuss "Personalized Medicine" or "Precision Medicine" as potential titles. Some might also wonder why not use the term "Genomic Medicine" to describe the contents of this endeavor. Given this potential controversy, or confusion, let's take the opportunity to better define what we understand by these concepts and their historical and cultural context. The first concept associated with the use of whole genome sequencing and the Human Genome in relation to health is that of Personalized Medicine. At the beginning of the twenty-first century, with the reduction in sequencing costs, several scientific and popular publications promoted the use of this term in direct relation to sequencing an individual's genome for diagnosis and treatment purposes. The underlying argument for using the term "Personalized Medicine" was that since the "Personal Genome" is sequenced, i.e., the genome of the individual, analyzing it would allow identifying their personal characteristics (or traits) that would lead to a personalized diagnosis and treatment. Although the concept was a blockbuster success, the medical establishment (mainly those outside the use of genomics) opposed to its use arguing that all medicine was personalized, as it always sought to treat the individual patient according to their personal characteristics. In this context, the use of personal genomic information represented an advance in this direction, but not a paradigm shift.

With the decline in the use of "Personalized Medicine", the term "Precision Medicine" began to gain strength. This concept in relation to genomics arises from developments in the second generation of oncology drugs, which unlike traditional chemotherapy (based, for example, on the use of DNA intercalators like cisplatin that kill all dividing cells), aimed to inhibit specific proteins of the tumor, derived from oncogenic mutations. These second-generation chemotherapeutics, mainly Kinase and Phosphatase inhibitors used in lung cancer and certain leukemias, are based on the knowledge of the mutations (variants) that the tumor possesses and therefore make it susceptible to them. In this context, the term Precision Medicine sought to mark the difference between general chemotherapy and that which is precisely based on the particular tumoral genome. From this conception, its use expanded to refer to any treatment based on the knowledge of an individual's genome, with the argument that this knowledge would allow for a more precise medicine. What happened with the concept of "Precision Medicine" was that other branches of medicine were also using it, mainly associated with an improvement in the precision of the patient data acquisition process. For example, the change from measuring a patient's blood pressure in the medical office, which generates a data point at each visit, was replaced by the use of a digital blood pressure monitor that measures pressure in real-time over a week. This was added to the use of advanced imaging techniques (MRIs, CT scans, etc.) that generated more and more data, and better data. This data, it was argued, allowed doctors to have a much more precise view of the patient, and therefore make better decisions. This broad view based on the massive use of data is the conception that is currently most used for the term "Precision Medicine". In this conception, Clinical Genomics is part of "Precision Medicine", as it uses massive genomic data, but the two are not synonymous.

Lastly, we can consider the term "Genomic Medicine," which I believe to be almost synonymous with Clinical Genomics, as both imply the application of genomics within medicine. Although "Genomic Medicine" aligns closely with other medical specialties and might seem more fitting, the distinction between Genomic Medicine and Clinical Genomics primarily lies in their focus. Genomic Medicine is more broadly associated with a specific field of medicine and therefore refers not only to its implementation but also to its underlying biological principles. On the other hand, "clinical Genomics" specifically emphasizes the translational aspects, focusing on direct clinical interventions in patient care. This emphasis is the one I like to promote in the present book.

The Future of Clinical Genomics

We could not conclude this introductory chapter without looking ahead to the future of Clinical Genomics. As a famous saying among scientists goes: "making predictions is difficult, especially about the future." Despite this, and based on the developments observed in recent years and the knowledge of various ongoing research projects, we can venture some interesting predictions. In the near future, we may see the advancement and culmination of various projects that are corollaries of the Human Genome Project (which we will detail in Chap. 3). Firstly, there is a clear trend toward Population Genomics projects that aim to understand with ever-greater detail and precision the variation of genomes within our species. A more ambitious project is the continuation of the ENCODE project, which aims to catalog all functional elements of our genome. In line with what we have seen in this chapter, this project aims to translate into the language of biological function all the elements (or segments) of the human genome sequence. An even more ambitious project is the so-called "Human Cell Atlas" which seeks to catalog the gene expression state (i.e., determine the expression level of each of the human genes, using MST) of every human cell type throughout the entire embryological process and in adults. This catalog will be an ideal complement to the Human Genome, as it would tell us which genes are active in each cell of each type of tissue. While these projects are based on using MSTs and may impact Clinical Genomics, they are general scientific projects that address fundamental areas of our biology.

In direct relation to Clinical Genomics, we can mention the following challenges for the near future. The first consists of cataloging and defining all causal variants of Mendelian pathologies (which we will detail in the corresponding chapter). Directly related to this goal, the near future will likely bring the development of better predictive models that allow for more accurate predictions of which new variants (i.e., not previously reported in another individual) are causal for a pathology. Furthermore, in the area of prevention, it is expected that the expansion of Genome Wide Association studies (GWAS) and the development of predictive models for individual risk of developing polygenic diseases will result in an increasing adoption of their use in medical practice. In relation to treatment, a growing adoption of pharmacogenomics is also anticipated. Lastly, there are various ongoing developments related to MSTs that we will see in the next chapter, which are expected to continue reducing costs and expanding accessibility to genomic information.

Regarding the more distant future, and therefore more difficult to predict, we can mention the following aspects. Firstly, it is clear that Artificial Intelligence will increasingly participate in Genomics in general, and Clinical Genomics in particular. On the one hand, it will have a clear impact on the development of better algorithms for the various steps of genomic data analysis, and why not the development of an "AI-based Genomic Doctor." An early glimpse of this type of technology was presented a few years ago by IBM, who developed Watson, an AI descendant of "Deep Blue" that beat then-world chess champion Gary Kasparov in a memorable chapter of the eternal battle between man and machine. Watson-IBM then won the famous TV game show "Jeopardy" of questions and answers and showed to be capable of diagnosing a patient with a genetic disease based on genomic data and reading numerous scientific articles. However, the project was discontinued, and today there is still no digital Dr. House based on AI.

Secondly, it is possible that we will see a growing approval of gene therapies, using the new methods of genetic editing based on the use of CRISPR-type ribonucleoproteins. These therapies aim to correct a genetic variant in the affected individual, typically through CRISPR-based systems. Such interventions require precise molecular diagnosis and raise significant ethical and regulatory considerations. In this sense, it is clear that Clinical Genomics and gene therapy are close sisters. The third aspect of the future is perhaps the most distant (or not?), the most controversial, and one that awakens numerous fears and fantasies. It is based on using our knowledge of human genomics and the mentioned tools (TseqM and gene editing) to modify either in-utero or in-vitro human embryos to do what is known as eugenics. To understand what we are talking about, we can turn to Hollywood, to a scene from the movie GATTACA (note the letters used in the name of the movie, which refers to an institution associated with space exploration in the movie itself). In that scene, two future parents meet a "future obstetrician" who tells them that the best combination of both parents' genes, something that has a one in a million probability, has been selected to generate the "genetically" best possible child. The obstetrician also mentions that all genes associated with diseases or undesirable personality characteristics have been removed. The parents wonder if it wouldn't be better to leave something "up to chance"... But I will not "spoil" how the movie continues and ends. While it represents a scene of science fiction, it is not difficult to imagine how the combination of sequencing the genomes of both parents (or the embryo) combined with techniques to manipulate the genome could result in a similar scenario. While a situation like the one described in GATTACA is still far from reality, current genomic techniques already present controversies and intense medical, ethical, legal, and social debates associated with how we have children. Some of these, we hope to address in the corresponding chapter.

Market Outlook

To conclude this chapter, let's briefly discuss economics, or more precisely, the market. It's interesting to note, briefly, as it will have a significant impact on the outcome of making Clinical Genomics a reality, the current market situation for Genomics. Like any new technology, we can segment the commercialization of new technology into three stages. In the first stage only a few players will have access to it (the early adopters), and it will be provided by pioneering companies, usually startups, or side projects of established companies. In this stage, profits for providers are usually low while costs for consumers are high. In the second stage, the adoption stage, the market grows exponentially as many participants join, both from the provider and consumer sides, creating a virtuous cycle of supply and demand that results in wider access to the technology, decreasing cost, and increasing profits for providers. This transition is typically driven by falling sequencing costs, increased competition, expanding clinical demand, and growing regulatory clarity, all of which make the technology more accessible and attractive. Finally, the market consolidates, relative growth decreases, and the Darwinian process results in a concentration of the offer. This improves providers' margins and usually results in better service for consumers. In this phase, economies of scale, standardization, and market selectivity stabilize the sector and favor a smaller number of dominant providers. When we analyze the market for Clinical Genomics within this framework, what we can observe is that for "Diagnosis of Mendelian diseases" and "Oncological Genomics" in key countries such as the USA, Europe, and Asia, the market is in a stage of consolidation with a trajectory of almost a decade. However, the application of Clinical Genomics in other areas, such as "Prevention of risk of polygenic diseases", is still in the early stages, and "Pharmacogenomics" is possibly in a transition from the early stage to the adoption stage. In contrast, in Latin America, diagnostic and oncological Clinical Genomics are in a stage of sustained adoption. This is the best stage for the implementation of a service of this type as it is the most dynamic stage. On the other hand, other branches of genomics are at the dawn of the early stage. In other words, currently in Latin America, it is perhaps the best time for the implementation and/or development of a Clinical Genomics service, and this book is your main ally. In the following chapters, we will dive deeper into the Human Genome and the technological revolutions that allow us to read it with unprecedented detail.

References and Suggested Further Readings

Watson JD, Crick FH. Molecular structure of nucleic acids; a structure for deoxyribose nucleic acid. Nature. 1953;171(4356):737–8. https://doi.org/10.1038/171737a0. PMID: 13054692.

Collins FS, Green ED, Guttmacher AE, Guyer MS. A vision for the future of genomics research. Nature. 2003;422(6934):835–47. https://doi.org/10.1038/nature01626. Epub 2003 Apr 14. PMID: 12695777.

Jameson JL, Longo DL. Precision medicine–personalized, problematic, and promising. N Engl J Med. 2015;372(23):2229–34. https://doi.org/10.1056/NEJMsb1503104. Epub 2015 May 27.

Pena SDJ, Tarazona-Santos E. Clinical genomics and precision medicine. Genet Mol Biol. 2022;45(3):e20220150.

For those who want a yearly update, Starting in 2019, the American Journal of Human Genetics (AJHG) has published an annual feature identifying ten key advances in applying genomic information to clinical care that were reported in the previous 12 months of published literature. See last years: Manolio TA, Narula J, Rupert A, Bult CJ, Chisholm RL, Ginsburg GS, Green ED, Hooker G, Jarvik GP, Mensah GA, Ramos EM, Roden DM, Rowley R, Taylor CO, Williams MS. Genomic medicine year in review: 2024. Am J Hum Genet. 2024;111(12):2585-8. https://doi.org/10.1016/j.ajhg.2024.11.002.

Chapter 2
(The) Human Genome(s)

Abstract This chapter examines the Human Genome through its historical, technological, and conceptual development, emphasizing its central role in modern Clinical Genomics. It begins with the Human Genome Project (HGP), highlighting its scientific vision, international collaboration, and the ethical framework established to guide genomic research. The chapter contrasts two major sequencing strategies: the hierarchical, map-based approach of the public consortium and the Whole Genome Shotgun (WGS) method pioneered by Craig Venter, whose success reshaped genome sequencing and bioinformatics. The competition between public and private initiatives is presented as a defining moment that accelerated progress and raised fundamental questions about data ownership and access. From a technical perspective, the chapter introduces key concepts in genome assembly, including reads, contigs, scaffolds, and the computational challenges of reconstructing large genomes. It then reframes the human genome as a dynamic reference map, continuously updated to better represent population diversity. Finally, the chapter explores genetic variation, emphasizing that individual genomes differ by millions of variants, including single nucleotide variants, indels, and structural changes such as copy number variants. These differences underpin human diversity and disease susceptibility, establishing the foundation for variant analysis in clinical genomics.

Keywords Human genome project · Whole genome shotgun · Genome assembly · Reference genome · Genetic variation

At this point, it is worth noting that the order of the next two chapters—"(The) Human Genome(s)" (Chap. 2) and "Reading DNA" (Chap. 3)—can be interchanged without disrupting the conceptual flow. Over many years of teaching and researching this subject, I have experimented with both sequences and found that each offers distinct advantages. Beginning with the structure and organization of the human genome provides a strong biological foundation, whereas starting with DNA sequencing

M. A. Marti, *Clinical Genomics*,
https://doi.org/10.1007/978-3-032-34667-4_2

technologies offers an immediate entry into the methods that made modern genomics possible. Whether you choose to start with the biological foundation of the genome or the technological methods of sequencing, both paths lead to a comprehensive understanding of Clinical Genomics and you may choose whichever order aligns best with your interests and learning style.

The Human Genome Project

The history of the Human Genome Project (HGP) is a remarkable narrative that blends scientific ambition, institutional rivalry, political strategy, and technological innovation—elements that ultimately shaped the foundations of modern clinical genomics. It illustrates the dynamic interplay between science, technology, and society. The project began to take form in the late 1980s, driven by advances in DNA manipulation and sequencing technologies (see Chap. 3). Its earliest conceptual roots emerged within the United States Department of Energy in 1986, motivated by the need to understand how radiation damages DNA, a pressing concern tied to nuclear energy safety in the post–World War II era. In 1988, after extensive debate, the U.S. Congress agreed to cofund the project alongside the National Institutes of Health (NIH), marking a major milestone in the evolution of genomics.

The HGP was officially launched in 1990, with a 15-year goal of determining approximately 95% of the human genome's sequence. James Watson was appointed its first director at the NIH. He had been instrumental in shaping the initiative, championing the idea of sequencing the human genome, advocating publicly for its importance, and defending the project in congressional hearings. Watson envisioned that a complete human genome sequence would transform our understanding of human biology and revolutionize the diagnosis and treatment of disease—anticipating the central tenets of Clinical Genomics long before the field had a name.

As director, Watson established several crucial scientific, political, and social frameworks that guided the project's development. He strongly believed the HGP should be an international public effort benefiting all humanity, and he successfully encouraged the participation of major research centers worldwide. Among these, the United Kingdom played a leading role under the direction of John Sulston, who founded the Sanger Centre (named in honor of Fred Sanger) with funding from the Wellcome Trust and the Medical Research Council.

Watson also recognized early on that the HGP would raise profound Ethical, Legal, and Social Issues (ELSI). Accordingly, he established one of the first dedicated ELSI programs within a major scientific endeavor. This group examined topics such as the privacy and confidentiality of genetic information, equitable access to the benefits of genomics, genetic discrimination, ethical considerations in human subject research, and the intellectual property implications of genomic discoveries. This proactive approach ensured that the project was conducted responsibly, ethically, and with a commitment to societal well-being.

A third essential aspect of Watson's leadership involved determining the scientific and technological strategy for sequencing the human genome—a topic explored in detail in the next section.

How to Sequence the First Human Genome

Although the public consortium had secured 15 years of funding, sequencing the human genome remained an immense technical challenge. The scientific strategy chosen to tackle this effort became one of the domains where James Watson left a lasting influence—and where many of the project's later controversies began. At the time, sequencing relied on the original Sanger method, which used parallel reactions with chain-terminating nucleotides and electrophoresis on polyacrylamide or agarose gels. Typical read lengths were around 500 base pairs, meaning that tens of millions of reads would be required to cover the human genome even at minimal depth. Moreover, no computational tools capable of assembling a genome of such size from fragmented reads existed. Aware of these challenges, Watson believed that the HGP should not be centralized at the NIH but instead distributed among many research centers. To achieve this, he designed a hierarchical sequencing strategy that enabled large-scale parallel work.

The strategy began by systematically fragmenting each of the 24 human chromosomes (22 autosomes plus X and Y) into segments of roughly 150 kilobases, which were then cloned into bacterial artificial chromosomes (BACs), a technology newly developed at the time. These BACs could be propagated in bacteria, generating large quantities of identical copies. Each BAC was then purified and further fragmented into smaller pieces of 1–2 kilobases, which were subsequently sequenced using the Sanger method. In parallel, the HGP aimed to identify enough "molecular markers" across the genome to construct a physical map, anchoring each BAC to known genomic positions.

Molecular markers are specific genomic sites that can be identified using straightforward molecular biology techniques such as PCR, restriction digestion, and gel electrophoresis. Using classical genetic mapping approaches, distances between markers can be estimated (in centiMorgans, where 1 cM is roughly equivalent to 1 megabase). By combining marker distances with their presence in individual BACs, researchers could determine their order and construct a physical map of each chromosome. This hierarchical mapping strategy enabled the assembly of a detailed genome-wide physical map and facilitated the identification of key disease-associated genes early in the project, including MSH2 in 1993 and BRCA2 in 1995.

Another key decision advanced by Watson, and later formalized at the 1996 Bermuda conference, was the commitment that all genomic data generated by the HGP would be released publicly without delay. This policy ensured unrestricted access to the sequencing data and promoted collaboration, reproducibility, and rapid scientific progress. Data were deposited daily into GenBank. With funding secured, a clear scientific strategy in place, and rapid progress across participating centers, the project appeared to be advancing steadily—although the most significant challenges were still to come, emerging from within the NIH itself.

The Emergence of Craig Venter

Craig Venter, always a passionate scientist from an early age, was educated at the prestigious University of California, San Diego, where he had the privilege of studying under the tutelage of renowned biochemist Nathan Kaplan. Here, he delved into the study of biochemistry, exploring the complexities of physiology and pharmacology, and he graduated in 1972. He began working as an associate professor at the State University of New York in Buffalo, where his talent and dedication soon led to his promotion to full professor, thereby solidifying his reputation as a brilliant and committed scientist. In 1984, he decided to join the NIH to research and study gene expression, focusing especially on the identification of mRNA and its role in regulating key biological processes. No one could anticipate that the twists of scientific research in this area would lead him to a direct collision with Watson and the HGP.

One of the "hot topics" in the study of mRNA at that time, which Venter actively worked on, was its transformation into ESTs (Expressed Sequence Tags). The technique involves using a Reverse Transcriptase (like that of the human AIDS-causing virus HIV) to convert mRNA into DNA (known as cDNA) that can then be analyzed using available molecular biology techniques, particularly Sanger sequencing. These ESTs represent, in some way, fragments of genes that are actively expressed at a specific time and a particular tissue, making them valuable tools for understanding gene activity and regulation in the cells. Moreover, their sequence usually contains the entire coding region, thereby allowing the corresponding protein sequence to be obtained.

In the early 1990s, while the HGP began its trajectory independently and unbeknownst to its nemesis, Venter decided to apply the EST technique to cultures of human neurons, obtaining the ESTs and sequences of about 2000 genes. At the NIH, before a study was published, it underwent an internal patentability review, and in this case, the NIH Office of Technology Transfer chose to explore the possibility of patenting the identified genes. This decision sparked intense controversy in the scientific community and the political arena, triggering a series of debates on the intellectual property of genes and the ethics behind the patentability of genetic elements, particularly those belonging to our species. The controversy provoked a negative reaction from the public and other researchers in genetics and biotechnology, which Watson took as a personal affront and against the HGP. Although internal discussions and the community's reaction led the NIH to withdraw its request to patent human genes, the controversy did not end there and reached the U.S. Supreme Court through other actors. Meanwhile, Venter's growing interest in genomics increased his visibility—but this was Watson's domain, and he was not willing to include the young and talented scientist in the HGP. Tensions surrounding this dispute contributed to Watson's resignation as HGP director, after which Francis Collins—a physician-geneticist known for discovering several Mendelian disease genes including CFTR—took over the leadership of the project.

Seeking to demonstrate to Collins and the public consortium that he and his group could make a significant contribution to the HGP, Venter had perhaps the most brilliant idea of his career, which led to the development of the strategy that is used today to sequence whole genomes, and which is known as "Whole Genome Shotgun

Sequencing." Venter had been successful in sequencing thousands of neuronal ESTs using the Sanger technique. In that case, each read or sequence corresponded to a different mRNA, and therefore they were "independent." But what would happen if, instead of sequencing all the mRNAs derived from a culture, one extracted the genomic DNA and fragmented it into segments similar in size to the mRNAs? Sequencing these "thousands" of fragments would be equivalent to obtaining the ESTs, only now all of them would be part of the same genome. If one could then reconstruct (assemble) the original genome from those fragments, the technique would allow for sequencing and assembling entire, or whole, genomes. Venter named this technique Whole Genome Shotgun (WGS), which later evolved and today is used for Whole Genome Sequencing. The name is related to the random spread pattern of a shotgun blast. In this case, the pellets expand from the shotgun's muzzle, forming a random pattern on the target. Venter "imagined" each of the DNA fragments as a random point in genomic space. Venter proposed, within the framework of the NIH, to use this technique to contribute to the HGP, but the public consortium did not believe the technique could work, due to the complexity involved in reconstructing the complete genome from the fragments.

As a last attempt to convince the public consortium of the capability and power of WGS, Venter decided to perform a proof of concept using *Haemophilus influenzae* as a model organism, a small pathogenic bacterium that can cause a variety of diseases in humans, including respiratory infections such as pneumonia, sinusitis, and otitis among others, especially in young children and people with weakened immune systems. Venter and his team sequenced approximately 20,000 fragments of about 2 kb, then assembled the complete genome of 1.8 Mb in length, corresponding to an average depth of 22×. The publication of the complete genome of *Haemophilus influenzae* in 1995 in the journal Science marks one of the major milestones in the history of genomics, laying the foundation for future large-scale genomic sequencing projects. However, despite the success, the HGP was reluctant to incorporate the technology, claiming that the complexity of a eukaryotic genome such as the human was not feasible to be sequenced by WGS. Reaching this point, Venter decided to leave the NIH and seek to sequence the human genome on his own, funded by a private company called Celera Genomics.

The Race for the Human Genome

In 1998, Celera Genomics, a private venture capital firm founded and led by Venter, was publicly announced. The company declared its goal to sequence the human genome before the public consortium, using the previously described WGS technique developed by Venter in the NIH. Celera intended to commercialize access to its genome data for other companies, research institutions, and organizations interested in using this information for various purposes, such as drug development (Big Pharma), biomedical research, and comparative genomics. Furthermore, it offered genomic sequencing and analysis services, and the development of bioinformatics tools to support these applications.

This event sparked a new conflict, bringing into the public arena fundamental questions about whether it is possible to patent genes, or who ultimately owns the heritage of our species genome. During the more than two years that the conflict lasted, the public consortium underwent intense internal debate. On one side were those who advocated for an aggressive approach, which led to starting the large-scale sequencing of BACs earlier than originally planned. On the other side were the extreme skeptics who did not believe that Celera and its WGS could actually sequence and assemble the whole human genome. The moment of revelation for the public consortium skeptics occurred in the year 2000 when Venter and Celera published the complete genome of *Drosophila melanogaster*, the fruit fly, one of the emblematic organisms in genetics, in the journal Science. The paper described in great detail the bioinformatics method developed to assemble the whole genome using short reads, which is about one quarter the size of the human genome. This work definitively erased any doubts about the power of the WGS technique, and the bioinformatics needed for the assembly of large eukaryotic genomes, such as the human genome. When this work was made public, the consortium members of the HGP were thrust into panic when forced to confront the real possibility of being overtaken.

At the beginning of the twenty-first century, the HGP thus faced a potential public relations nightmare, since if Celera "won" the race, they would need to explain why they had spent billions of dollars in taxpayer money and lost to a small group of entrepreneurs who "supposedly" possessed an unworkable technique. The issue was compounded by Venter's personal animosity towards the NIH arising from his earlier conflict with Watson, and his rejection as a participant in the HGP, which was difficult to navigate at the time. Given the situation and the potential negative impact on the scientific community, someone eventually managed to alert the highest authority to take action. As a consequence, at that time, the President of the United States, Bill Clinton, who was a friend of both Venter and Collins, tasked the Secretary of Science with finding a way for "both" parties to reach an agreement. After many discussions and negotiations, an agreement was finally reached in June 2000, culminating in an international press conference where Bill Clinton and the UK Prime Minister Tony Blair, along with Francis Collins and Craig Venter, announced the sequence of the first draft of the Human Genome, in an iconic photo that anyone can now view online. Clinton referred to the event as the "discovery of the century—perhaps of all time." Thanks, or as a consequence of Venter, the human genome had been mapped almost five years earlier than originally planned. Three years later, the project was officially completed, with the joint publication of the complete genome sequence in the prestigious journals Science (by Venter's group at Celera) and Nature (by Collins and the public consortium).

The controversy over the motives behind creating a private initiative to sequence the human genome, in direct competition with the public consortium, reveals two opposing approaches in scientific and technological production. On one hand, there is state-funded research aimed at contributing to the general welfare, and on the other, the pursuit of knowledge driven by private investors for commercial purposes. This can also be likened to the old and well-known conflict between basic science and applied science, or between science and technology. These contrasts pose a significant challenge in today's society: to find a balance and establish regulations to harmonize these divergent approaches, a challenge that also manifests in the field of genomics.

Yet the question remains: what were the real motives behind Celera's objective? Venter claims that his motivation for sequencing the genome was primarily and purely scientific and technical, not commercial. That is to say, fundamentally, he believes he wanted to demonstrate that sequencing the human genome via WGS was possible and more efficient than the hierarchical strategy designed by Watson. However, some critics argue that Celera's method succeeded only because it used public information, such as knowledge of the physical map of molecular markers and even many publicly available sequences derived from the HGP. Others believe that the true motivation behind the genome race was commercial, given that Celera at that time controlled the sequencing equipment market, which gave them a significant competitive advantage. It is interesting to note that one of Celera's main financiers was ABI, which sold Sanger-type sequencers and was in the midst of developing capillary sequencing technology with fluorescent nucleotides, and that Celera served as a "test bank" for this equipment. Moreover, the genome race resulted in the public consortium acquiring thousands of sequencers, which represented a notable influx of funds for ABI. However, it is also important to note that Celera's business model utterly failed. This is not surprising since it is hard to imagine pharmaceutical companies paying for information that would become public (that is, the sequence of the human genome) just a few weeks, or months later. For its part, a year after publishing the genome, Venter resigned as CEO and Celera shifted into a pharmaceutical development company, ceasing its research in genomics. Surprisingly, ABI failed to capitalize adequately on its gains and lost the market race when second-generation sequencers, described in Chap. 3, appeared and Schumpeterized the sequencing market a few years later.

Finally, we can reflect briefly by putting the HGP in perspective. At the time of its publication, in a significant statement, Collins asserted that the complete sequencing of the human genome represented an unprecedented opportunity to understand the role of genetic factors in health and disease, and to apply that knowledge in prevention, diagnosis, and medical treatment. As we have seen, these are the pillars of Clinical Genomics that we all seek to make a current reality. Additionally, Collins made a bold prediction indicating that by 2020, knowledge of hereditary factors associated with common diseases would allow for the early prevention of these conditions. He also anticipated significant advances driven by genomics, including the development of personalized drugs for diseases such as diabetes, tailored therapies to combat cancer, and a revolution in the treatment of neurodegenerative diseases. Today, 20 years later, we can see that while many of these predictions may have taken slightly longer than expected, they have indeed become reality. However, translating these ideas into reality is a laborious task. As often happens with many groundbreaking discoveries, we, as a society, tend to overestimate their short-term impact, leading many to become detractors of genomics, only to underestimate their long-term effects. Currently, it is possible that the community is not yet fully aware of the change that the HGP, and consequently, Clinical Genomics represent for research and, above all, medical practice.

The different Strategies and Their Impact on the Human Genome Assembly

In this section, after having presented and analyzed the sociopolitical debate associated with the HGP, let us examine both strategies from a methodological or technical perspective and see how Celera ultimately managed to assemble the genome from the reads. To understand the complexity difference between the two strategies—the one used by the public consortium and that by Celera's—let's again use the analogy tool, and along the way, delve into a bit of genomic numerology. Remember that, ultimately, to obtain a genome, one must first obtain the reads and then assemble them to form the whole genome. Assembling the reads involves organizing them (i.e., determine which one comes first, which one follows, etc.), using the fact that contiguous reads usually have some part of the sequence in common, known as an overlap.

As an illustrative example, let's consider the following two reads: (1) ATGTACTACTGACTACGTTACGCCATCATCAGCATCA and (2) ACGCCATCATCAGCATCATGCATCACGTACATCGATGCTATTCTTA

As can be seen, the end of read 1 (3′ end) matches the beginning of read 2, allowing them to overlap in this way: ATGTACTACTGACTACGTTACGCCATCATCAGCATCA ACGCCATCATCAGCATCATGCATCACGTACATCGATGCTATTC

If we now combine both reads, the following sequence results (bold indicates where both reads overlap): ATGTACTACTGACTACGTT**ACGCCATCATCAGCATCA**-TGCATCACGTACATCGATGCTATTCT

In this way, we can overlap several reads, as shown below: ACTGAT CTGSTCTT TCTGSTCTTGSTASASTGTTGAC GTTGACSTGACCTG CCTGATGCTCTGAGATTAAGGA TAAGGAGCCTAGGG And determine the sequence resulting from combining them all: ACTGATCTGSTCTTGSTASASTGTT GACSTGACCTGATGCTCTGAGATTAAGGAGCCTAGGG

The sequences that result from ordering the reads based on their overlaps, resulting in a "continuous" sequence without gaps, are called CONTIGs. Once sequenced and assembled, genomes end up being an ordered collection of CONTIGs, since only for small genomes is it possible to know the entire genome from end to end, or in other words, obtaining a single CONTIG.

Going back to our analogy, we can think of each "read" as a piece of a linear puzzle, where each piece joins to two others (the joining is done by the read overlaps). The strategy of the public consortium consists of taking the entire genome, which measures about 1 Gb (1 billion bases), and fragmenting it into about 1000 segments, each forming a BAC of 1 Mbp (1 million bp) in length. Each BAC is then fragmented, and the fragments are sequenced by the Sanger method, resulting in about 1000 reads of 1000 bp in length. Thus, assembling each BAC is equivalent to completing a 1000-piece (linear) puzzle. To assemble the entire genome, one must assemble the thousand 1000-piece puzzles (one for each BAC), and then put together a final definitive puzzle of a thousand pieces, where now the larger pieces are the BACs themselves. It's important to highlight that the first thousand puzzles can be assembled in parallel by different individuals. Since completing a 1000-piece puzzle is perfectly feasible, it is clear that the public consortium's strategy had a guaranteed chance of success. The human genome of the public consortium, published in Nature, using this strategy had an average depth of 1.5× and covered 92% of the human genome.

Now let's consider Celera's approach. Celera proposed fragmenting the entire human genome into millions of fragments and, bypassing the BACs, going directly to the reads. Using numbers similar to the previous example, this would mean fragmenting the 1 Gb genome into 1M (one million) reads, each one thousand bases long. So, what would the puzzle analogy look like here? Assembling the complete genome from one million reads is like assembling a puzzle of one million pieces. While I'm not sure what the Guinness World Record for the world's largest puzzle is (Google suggests it's one from Kodak with 51,300 pieces, this is far from one million), it is clear that the challenge of assembling such a large puzzle is deemed very difficult, if not impossible (even for a computer). To solve this issue, Celera acquired what was at that time one of the world's most powerful available supercomputers, and developed very powerful bioinformatics algorithms, whose fundamental ideas are still used today to assemble genomes. These algorithms are based on the construction of graphs, which efficiently separate the reads into future Contigs, and assemble the corresponding puzzles piece by piece. In a second stage, the algorithms use the information associated with "pair-ends" to order the contigs into higher units called scaffolds. The fact is that Celera solved the million-piece puzzle problem and obtained a complete human genome directly from the reads, demonstrating that the WGS technique was viable. However, as a final affront to the public consortium, in addition to publishing the genome obtained directly from the reads, Celera published a second genome, which was of significantly higher quality than that obtained by the public consortium, taking advantage of all the information available at the time and using a combined and ingenious strategy. Let's now see how Celera reached this best possible human genome.

To obtain the best possible human genome, Venter and his team at Celera Genomics decided to combine their assembly strategy. On the one hand, Celera sequenced 27M reads of approximately 550 bp. On the other hand, they took the information available up to that moment in GenBank which were derived from the public consortium. This information included the preliminary sequence of several chromosomes, from the assembly of their corresponding BACs, and the published information of molecular markers. The sequence information from GenBank comprised about 800 Mbp in total and with it, Celera built 16M "false" (or simulated) reads, which they added to the reads obtained from their sequencing. In total, the sum results in an expected depth of 5× far superior to that of the public consortium. To facilitate assembly (solving the puzzle) they compared all reads with the known sequences of the BACs, determining which reads corresponded to a given BAC and which were loose reads. Then, similar to what the public consortium did, they generated the contigs for each BAC. With the loose reads, they applied their assembly algorithm and generated an additional set of contigs. Third, they used the pair-ends information, along with the molecular marker data, to order the contigs and obtain the final genome. This genome covered about 96% of the human genome and had fewer and smaller gaps than the genome of the public consortium. However, Venter and Celera's use of public information (BACs and Molecular Markers), was very poorly received by the scientific community. Before moving on to understand what represents the sequence of the human genome and how it is used in clinical genomics, the final question that remains for you, the reader, in this context is: If you were in Craig Venter's shoes, would you have used the available information to obtain "the best" possible genome, or would you have published "only" the genome obtained with your own data?

The Human Genome as a Map

The human genome, whichever of the two that was published, ultimately amounts to a sequence of letters (or bases) starting on Chromosome 1, and just after about 10 kbp undefined letters, with 5′-TAACCCTAACCC... and continuing like this up to the last base of chromosome Y. Indeed, as we already mentioned, it corresponds to a collection of ordered contigs—for example, the first contig that one can view in RefSeq is NC_000001.11, indicating it is the 11th version of the contig of Chromosome 1. Currently, the sequence of "The" human genome is known as the Human Reference Genome (HRG), which is updated approximately every six months by the Genome Research Consortium. This is why the different human reference genomes used over time are known as GRCh1 (human 1), GRCh2, and subsequent versions. Currently, at the time of writing this, we are using GRCh40.

The HRG initially provides us with a one-dimensional map. Unlike a location map, such as the one shown by Google Maps, which features two coordinates (latitude and longitude), the HRG has a single coordinate defined as Chromosome:position. So, for example, 1:2,543,334 corresponds to position two million five hundred forty three thousand, three hundred thirty-four on Chromosome 1, and if we search, it will tell us that this position corresponds to a "T." Similar to Google Maps, the HRG also adds two features. On one hand, it has different "layers," equivalent to the Terrain/Satellite/Traffic views on Google Maps, which in our case corresponds to the different layers of biological information (where the genes are, the regulatory zones, introns, exons, their reading frames, and thus to which protein residue they correspond, as well as the known/reported variants of each position). This information is known as the genome's structure. And, analogous to Google Maps, the HRG allows for zooming in and out, viewing from the structure of the chromosomes and chromosomal banding (minimal resolution) to the sequence of individual bases (maximum detail), passing through all intermediate levels that allow visualizing the relative position of the genes, their extension, location, internal structure, and other features. In summary, the HRG is a map of all the positions in our genome, and everything we know about each region at the molecular level and beyond.

Two frequently asked questions concerning the HRG are, "Whose is it?" and "How does it evolve?" Both questions are interrelated and are deeply connected with the history of the HGP. In the public consortium, samples for sequencing the HG were obtained from volunteer donors and a "pool" of different randomly selected individuals was created, additionally representing different races/ethnicities according to the classification used in the USA (European, African-American, Asian, Hispanic, etc.).

On the other hand, Celera used a pool of samples taken from Venter and other company managers. It's clear that with this source, the initial HGs had a significant sampling bias, with entire regions that "by chance" corresponded to the genome of "one of the original individuals" and therefore did not adequately represent the diversity of our population. Subsequent updates to the HRG seek to address this initial sampling bias by replacing the reference sequence with those that are more representative. Ultimately, the updates to the HRG aim, to "close" the gaps between the contigs, and also to generate a sequence that is more representative of the general population. The first of these goals has been successful, and last year (beginning of 2023) the HG was completely closed. This implies, theoretically, that each of the chromosomes is formed by a single contiguous sequence without gaps. The second goal is much more complicated, as it is difficult to determine what the most representative HRG would be. Currently, there are numerous population genomics projects that allow for studying genomic variation much more efficiently and in-depth (see Chap. 10). To conclude this topic, we can draw our attention to the following fact. The HRG tends to present a genome that, at each position, displays a base that is highly represented in the general population (with a greater tendency to coincide with the Caucasian European population due to historical reasons). It is from this reference, and as we will see next, that the genomes of each individual are defined and used in Clinical Genomics.

We Are All Mutants

To end our journey on the human genome, we turn our attention to the inherent diversity found within the genetic blueprints of individuals. This diversity not only underscores the uniqueness of each human being but also presents critical insights into how genetic variations can influence everything from physical traits to susceptibility to diseases. Anyone who takes a walk through their neighborhood will notice that all humans look different. Moreover, the differences observable with the naked eye are just the surface; internally, we are also distinct. The reason we are all different is that each of our genomes is unique. While we all belong to the species *Homo sapiens sapiens* and share the same architecture of "The" human genome, the genomes of each individual present variations. Technically, these differences are called variants. Variants are thus differences that a given individual genome presents compared to the Human Reference Genome. The most common variants are Single Nucleotide Variants (SNVs), which involve the change of a single nucleotide base (or letter) at a specific position between the individual's genome and the reference genome.

For example, if at position 4476 on chromosome 21 of the HRG there is a base T, but in an individual a base G is observed, this is recorded as an SNV and denoted as Chr21 Pos4476 T>G. Another type of variants are insertions or deletions of a few bases (about 5–20) relative to the HRG, known as small indels. For instance, if at position 3327 on chromosome 7 the reference human genome reads AACTTG, but in another genome reads AATG, this indicates a deletion of two bases and is denoted as Chr7 Pos3327 CT=>"".

Beyond these variants, relatively small in relation to the size of a chromosome, there are so-called structural rearrangements that involve variants in large segments of sequences from one (or more) chromosomes. Among these rearrangements, which can involve the movement of a chromosome section to another location or chromosome (translocation), the most common are those that result in an alteration in the zygosity of the segment. This occurs due to errors in meiosis. This change in zygosity results in a variation in the number of alleles of the whole segment and therefore the number of copies of the region and all the genes it contains, which is why these variants are known as Copy Number Variants (CNVs). A well-known example of a CNV is trisomy of chromosome 21, which involves the presence of three copies of this chromosome instead of the usual two and is responsible for Down Syndrome. In other words, while the genome of an individual consists of a sequence of bases, it is worked with, and understood, as a list of differences, called variants, relative to the HRG.

But how many variants does a genome typically possess? A particular genome exhibits between two and five million variants (the vast majority being SNVs and indels). Of these millions of variants, about 100,000 correspond to coding regions (i.e., the exome) and between 15–25,000 are usually variants that change the amino acid of the resulting protein, hence they are missense variants. These numbers are significant because they give us an idea of the magnitude of the work involved in processing and analyzing the results of a genomic study.

To visualize the complex hierarchical structure and the resulting diversity of the unique maps that make up each person, see Fig. 2.1: The Human Genome: The Blueprint as a Living Map.

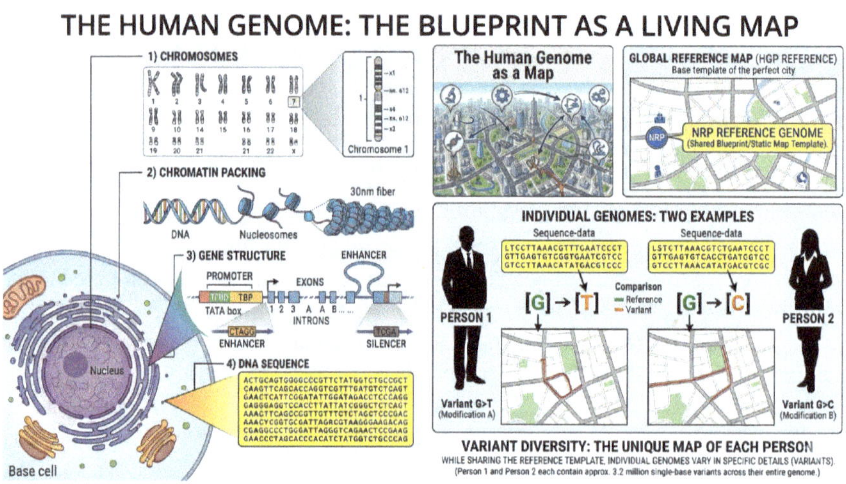

Fig. 2.1 The human genome as a living map. Hierarchical organization of human genome, including detailed gene structure and the underlying nucleotide sequence. Human genome as a map analogy, highliting individual-specific variation, exemplified by distinct single-nucleotide polymorphisms and sequence modifications for two individuals

Having thus understood that the genome of a patient will be in our hands a list of variants, in the following chapters we will learn how, when the genome (or a fraction of it, such as the exome) of a patient is sequenced and the reads are obtained (Chap. 3), they are used to efficiently and accurately determine the variants present in the individual's genome. This process is known as "variant calling." But before going into the variants, we have to learn how genomes are sequenced and what are the basic bioinformatic tools that allow us to process and analyze them.

References and Suggested Further Readings

Hood L, Rowen L. The human genome project: big science transforms biology and medicine. Genome Med. 2013;5(9):79. https://doi.org/10.1186/gm483. eCollection 2013.

Venter JC. A life decoded: my genome, my life. Viking Press; 2007.

Donohue CR, Alan CL. Perspectives on the human genome project and genomics. University of Minnesota Press; 2026.

McElheny VK (2010). Drawing the map of life: inside the human genome project. Basic Books.

Fleischmann RD, et al Whole-genome random sequencing and assembly of Haemophilus influenzae Rd. Science. 1995;269(5223):496–512. https://doi.org/10.1126/science.7542800.

Adams MD, et al The genome sequence of Drosophila melanogaster. Science. 2000;287(5461): 2185–95. https://doi.org/10.1126/science.287.5461.2185.

Venter JC, et al (The Celera team). The sequence of the human genome. Science. 2001;291(5507):1304–51.

Lander ES, et al (International Human Genome Sequencing Consortium). Initial sequencing and analysis of the human genome. Nature. 2001;409(6822):860–921. https://doi.org/10.1038/35057062.

Chapter 3
DNA Sequencing

Abstract This chapter presents the evolution and principles of DNA sequencing technologies, highlighting their central role in enabling modern genomics. It begins with the foundational Sanger method, which established sequencing by synthesis using chain-terminating nucleotides and remains the gold standard for accuracy. The chapter then introduces the polymerase chain reaction (PCR), a key technique for DNA amplification that underpins many sequencing workflows. The core focus is on massive sequencing technologies (MST), also known as next-generation sequencing, which transformed the field through miniaturization and parallelization, dramatically reducing cost and time while generating millions of short reads. Major second-generation platforms, including pyrosequencing, Ion Torrent, Illumina, and MGI, are described in terms of their technical principles, strengths, and limitations. Key concepts such as reads, pair-end sequencing, and library preparation are introduced, providing the basis for experimental design in genomics. The chapter also explores third-generation sequencing technologies, including PacBio and Oxford Nanopore, which enable long-read and single-molecule sequencing, offering new advantages alongside higher error rates. Finally, it discusses future directions, emphasizing the pursuit of direct, real-time DNA reading and continued cost reduction, reinforcing sequencing technologies as a driving force behind the expansion of clinical genomics and its applications.

Keywords DNA sequencing · Sanger method · Next-generation sequencing · Reads · Nanopore sequencing

DNA Sequencing

Since the structure of DNA was uncovered in 1953 by Watson and Crick, and the scientific community understood that the information is stored in the order, or sequence, of the bases on a DNA strand, there has been a continual development

M. A. Marti, *Clinical Genomics*,
https://doi.org/10.1007/978-3-032-34667-4_3

of more and better technologies for manipulating and sequencing DNA. At the start of the twenty-first century, as a result of the Human Genome Project, these technologies made a "qualitative" leap in their ability to read DNA, ushering in the genomic era. What are these technological changes that permitted this leap? What do we understand by "next-generation" sequencing, or massive sequencing technologies? What (bio)technological opportunities do these technologies offer? How is their development moving forward, and what can we expect in the near and long-term future? To answer these questions, we will explore the history of sequencing technologies, focusing on the key technological ideas and advances of each era, seeking to understand the technical and conceptual steps of the underlying processes, and how these impact the capabilities, strengths, weaknesses, and applications of each technology and their variants. Finally, we will introduce key concepts, such as reads, the exome, capture, etc., which we will use in the context of clinical genomics.

A Nobel Prize? Please, Give Me Two!

Winning the Nobel Prize is undoubtedly the pinnacle of any scientific career. An extraordinary recognition both from peers and the global scientific community. Achieving it is a colossal feat, and even Albert Einstein, the prototype genius scientist of our era, received it only once, for explaining the photoelectric effect, and not for his more famous theory of relativity. Only four scientists have achieved this distinction twice in more than a century: Marie Curie (in Physics and Chemistry), John Bardeen (both in Physics), Linus Pauling (in Chemistry and Peace), and Fred Sanger. In the early 50s, Sanger deciphered the complete sequence of amino acids in insulin, a small protein produced by the pancreas that is essential for cellular glucose absorption, whose deficiency is typical of type 1 diabetes. This discovery, which showed that proteins have specific, and unique, amino acid sequences, earned him the Nobel Prize in Chemistry in 1958, and laid the groundwork for later determining the genetic code. After this breakthrough, Sanger focused his interest on nucleic acids, specifically DNA.

In 1967, Sanger began to explore how to sequence DNA, a process that would, essentially, allow reading a whole genome. Employing DNA polymerase, essential in DNA duplication, radioactive nucleotides, and a separation technique, he managed to sequence the first complete genome in 1975. That of the bacteriophage φX174 virus, which contains 5386 bases (or about 5Kb). Two years later, he introduced what is now known as the "Sanger method," a revolutionary advance in DNA sequencing that earned him his second Nobel Prize, again in Chemistry in 1980. This method was key in the Human Genome Project, which was completed in 2001. Fred Sanger passed away on November 19, 2013. His findings transformed the world, yet his legacy often appears understated only because we have not yet fully appreciated his magnitude. Now, let's examine the "Sanger method" in detail.

Sanger's Method for DNA Sequencing

Sequencing a DNA molecule, in theory, seems straightforward. It starts at the 5′ end and identifies the first base or nucleotide, for example, an A. Then, we move on to the next base, let's say C, and so on until reaching the 3′ end. Thus, determining the complete sequence, such as ACTATCG. As we can see, sequencing a linear polymer involves identifying each of the monomers "in order." In other words, reading a sequence consists of determining in precise order (i.e., one by one) which letter of the alphabet appears at each position. Indeed, everyone learns to read their language in this way. Observing how a little girl (or boy) learns to read can make this clear. In summary, to sequence DNA, we must determine the order and nature of the bases, one at a time.

Sanger and his team used electrophoresis to sort the DNA fragments. The electrophoretic techniques already in use at that time allow for separating molecules by their size. In the case of DNA, under the right conditions, this allows for separating single-strand DNA according to its length. With this technique at hand, Sanger reasoned in the following manner: Given a single-strand of DNA sequence, for example, 5′-ACTACTG featuring seven nucleotides. If it were possible to generate all fragments of different lengths starting from the 5′ end, the following set of molecules would be obtained: A(1), AC(2), ACT(3), ACTA(4), ACTAT(5), ACTATC(6), and ACTATCG(7), which could be sorted by electrophoresis from largest to smallest, yielding the following result:

ACTATCG ACTATC ACTAT ACTA ACT AC A

It is important to note, that if it were possible to identify the last base of each fragment, this would reveal the desired sequence 5′-ACTATCG-3′. Thus, according to this idea, what is needed to sequence a DNA strand, is a technique that allows, given a sequence, to generate all the fragments that start at the same point (the 5′ end), and subsequently to identify the last base in each of them.

Sanger's geniality consisted first, in using the enzyme DNA polymerase to generate the fragments. This enzyme creates a complementary strand from a template sequence and an appropriate primer (a short DNA oligonucleotide complementary to the strand whose complementary strand is to be synthesized). By using a single primer, Sanger ensured that all fragments started at the same point, the primer itself. The problem is that polymerases have a high processivity, meaning it produces long fragments until it finishes, and what Sanger needed was to generate all the fragments starting from the primer, plus a single base (5′-primer-A in our example). Additionally, he needed to identify what was the last base of each fragment created by the polymerase.

To solve these problems, Sanger relied on his knowledge and ability as a chemist and his understanding of the reaction mechanism by which new nucleotides are incorporated by the polymerase into an existing DNA strand. When the polymerase incorporates a nucleotide, a phosphodiester bond forms between the free hydroxyl group at the 3′ position (3′-OH) of the sugar of the last nucleotide on the strand, and the alpha-phosphate of the incoming triphosphate nucleotide. Sanger reasoned that if he provided the polymerase with dideoxynucleotides, i.e., nucleotides that in their ribose lack a 3′-OH group but are identical in the rest of their structure to those used by the enzyme, these would be incorporated into a strand, leaving, after their incorporation, an end where there is no 3′-OH hydroxyl, which thus prevents further nucleotide additions. Because of this ability of these dideoxynucleotides (those lacking 3′-OH) to terminate chain synthesis by the polymerase, they are called terminator nucleotides.

If in a reaction medium, a template strand, primer, polymerases, the four nucleotides (A,C,T, and G) triphosphates, and between 10% and 20% of dideoxyT are added, the polymerase will generate all the possible fragments that start at the primer and end in a T. In our example sequence, 5′-ACTACTG, the polymerase will generate two fragments: ACT and ACTAT. In the original method proposed by Sanger, four separate experiments were conducted, in which everything required for the reactions was added, and additionally one of the four terminator nucleotides. Subsequently, the four experiments were run on the same electrophoresis gel in separate lanes, and by determining which lane corresponds to each fragment, the complete sequence is obtained.

During the 1990s, the method was automated and improved, maintaining itself as the standard for DNA sequencing quality. The most significant advancement was the incorporation of fluorescence-labeled nucleotides, which emit fluorescence of a specific color when excited. This allows marking each of the four terminator nucleotides with a different color, enabling performing a single experiment with the 4 terminator nucleotides in the same reaction medium, and identifying the last base incorporated by its corresponding color. The second advancement consisted of replacing gel electrophoresis with capillary electrophoresis. Currently, the Sanger method through capillary electrophoresis and identification of the base incorporated by the color of its fluorescence comprises the "gold standard" of DNA sequencing methods.

The Sanger method allows sequencing a DNA fragment ranging from 0.5 to 1.5 kb. Each sequenced fragment resulting from a sequencing reaction is called a "read" (guess why?). Currently, sequencing equipment using capillary electrophoresis can perform between 4 and 32 sequencing reactions in parallel, so it is possible to obtain up to 32 reads (sequenced fragments) in a couple of hours.

Finally, it is important to emphasize that Sanger sequencing is based on what is called "sequencing by synthesis," since to sequence, the polymerase must synthesize the fragments. The cost of the method, therefore, mainly lies in the reagents required for the synthesis: polymerase, nucleotides, and fluorescently labeled terminator nucleotides. As we will see next, massive sequencing techniques dramatically reduce costs, mainly by using smaller amounts of reagents. But before seeing how they do it, let's briefly review of another fundamental technique necessary for manipulating and sequencing DNA, the polymerase chain reaction.

The Polymerase Chain Reaction (PCR) and Other DNA copying Methods

The development of the polymerase chain reaction (PCR), one of the simplest and most revolutionary advances in molecular biology, can be credited to the inspiration of Kary Bank Mullis in the early 1980s. Working at Cetus Corporation, one of the first biotechnological startups, where he was looking to identify specific mutations within human genomes. The challenge was to isolate and multiply a small genetic region that contained the gene of interest out of the entire genome (does this problem sound familiar?). According to Mullis, the idea for PCR came to him while driving through California in 1982, when he pondered how to use synthetic fragments as primers to replicate the desired DNA segment in vitro using a polymerase (similar to the Sanger method). Mullis quickly realized that if, in each cycle of the polymerase reaction, "only" the desired fragment was to be copied, then, after a dozen cycles, he would obtain millions of copies of the same fragment, while the quantity of any other region would remain constant. This revelation promised to solve two major challenges in DNA biochemistry: abundance and selectivity. Ten years later, Mullis received the Nobel Prize, and in 1991, Cetus sold the PCR rights to Roche for an astonishing 300 million dollars. Today, this technique is a cornerstone in far reaching fields such as molecular biology, genetics, forensic analysis, and above all, clinical genomics.

The PCR procedure starts with a sample of double-stranded DNA, whose strands are separated by heating to 90°C. Subsequently, additions are made: the polymerase, nucleotides, and specific primers are added. By reducing the temperature to about 70°C, the primers anneal correctly to their respective complementary sequence positions (annealing process). This enables the polymerase to begin DNA synthesis. Each strand serves as a template to synthesize its complementary, thus doubling the DNA of interest in each cycle. Repeating the process, the amount of target DNA grows exponentially. For example, starting from a single molecule, a billion copies are produced after just 30 cycles. If there are specific primers for a gene in a human DNA sample, PCR allows for the strategic selective amplification of that gene, facilitating its subsequent analysis, for example by using Sanger sequencing.

The primary technological advantage of PCR is that it allows us to obtain millions of copies of the DNA fragment of interest, which can then be subsequently used in various applications that extend beyond sequencing. In this context, it is pertinent to briefly mention some alternatives available that also allow achieving this goal. One such alternative, which was employed by Sanger to obtain many copies of the DNA he used to develop his sequencing method, involves using viruses, typically phages that infect bacteria, as DNA photocopiers. When a virus infects a bacterium or eukaryotic cell, it generates many copies of itself during the viral replication process. This process involves producing millions of viral particles, which not only will contain both the protein capsid of the virus, but also its genome. Therefore, if the viral genome is replaced with the DNA of interest and the bacterium (or cell) is induced to replicate it, this will result in millions of viral particles containing the DNA of interest being produced. Other notable methods include the use of plasmids and bacterial artificial chromosomes (BACs). Plasmids, small circular DNA molecules found in bacteria, can be engineered to carry and replicate specific DNA fragments within bacterial cells. BACs, on the other hand, are larger and can carry more substantial DNA sequences, making them ideal for cloning bigger genes or larger genetic elements. As will be shown in the next chapter, BACs were essential to the Human Genome Project. Both plasmids and BACs offer robust and reliable means for multiplying DNA, providing valuable alternatives for various genetic studies and applications.

Massive Sequencing Technologies (MST)

Massive sequencing technologies (MST), which emerged at the start of the twenty-first century, specifically starting distinctly in 2005, heralded a paradigm shift in sequencing methods. They were originally referred to as next-generation sequencing techniques (NGS). The objective of these technologies was to develop sequencers that were not only more efficient, but also more economical in terms of cost per sequenced base. The competition among various companies to develop them over the last two decades has been a thrilling race involving technological innovation, market strategies, corporate acquisitions, and bankruptcies. A tough competition that continues to be highly relevant today. The most significant outcome of this endeavor has been the dramatic reduction both in the cost and time required to sequence a human genome. The cost fell dramatically from approximately one million dollars and nearly a year in 2008, to 100 dollars and merely a few hours in the present day.

A MST device functions by conducting multiple sequencing experiments in parallel in very small volumes, and where all (or most) of the steps are automated. The reduction in the volume of each sequencing reaction allows for the use of significantly fewer reagents, resulting in a decrease in the average sequencing cost per base read. In MST devices, each process results in the sequencing of a fragment from the original sample, and each of these obtained sequences is termed a read. In other words, a sequencing experiment using MST yields a set of reads (ranging from thousands to millions), hence its massive scale. In the following paragraphs, we will specifically outline the methodology used by the leading sequencers available on the market with these technologies. We begin with the so-called Second-Generation Sequencers (the Sanger Method is often termed as first generation).

One of the First Commercial Massive Sequencers (now discontinued) was Roche's Pyrosequencer. The initial steps on this platform involves appending known sequences, termed adapters, to both ends of the DNA fragments that are targeted for sequencing. These adapters (which, as we will explain, are device-specific yet universally utilized) enable the equipment to recognize the target fragments and typically also act as the region for primer annealing. In Roche's pyrosequencing method the DNA fragments are combined with agarose beads that possess probes complementary to the adapters, ensuring that each bead, on average, captures a single target fragment. The beads are mixed with a reaction medium-oil blend, which isolates each bead in a microdrop of water within the oil matrix. Each microdrop is equipped with polymerase, nucleotides, and appropriate primers. Subsequently, PCR cycles are conducted in the water-oil emulsion, where each droplet serves as an independent PCR reaction thus producing millions of copies of the bead-bound DNA fragment. At the end of this process, each bead is enveloped by millions of copies of the same DNA fragment. This procedure is usually referred to as clonal amplification, since each bead contains millions of clones from one of the original target fragments.

The beads with the amplified DNA are subsequently deposited into a matrix filled with numerous tiny wells, each designed to hold just one bead. Sequencing by synthesis is then performed on this bead array sequentially. In each "cycle" of sequencing, the polymerase and only one of the four nucleotides (A, C, T, or G) are introduced, followed by the next nucleotide in a subsequent cycle. The third nucleotide, then the fourth, and then the four step-by-step nucleotide cycle begins anew. The method of detection that identifies which wells incorporated a specific nucleotide (i.e., a specifi DNA base or letter) is based on a reaction that senses the presence of the released pyrophosphate upon polymerase nucleotide addition (hence the name, pyrosequencing). This reaction utilizes an enzyme known as Sulfate Adenyl Transferase, which swaps the sulfate group for the pyrophosphate in its substrate Adenyl-sulfate, therefore producing sulfate and ATP. This ATP is then utilized by another enzyme, luciferase, to catalyze the oxidation of luciferin, resulting in light production. The intensity of light generated is proportional to the volume of pyrophosphate released, and consequently, to the number of nucleotides integrated into that well during that cycle. Each well chronicles which nucleotide was incorporated during each cycle, indicating the sequential inclusion of nucleotides and facilitating the synthesis of the DNA sequence. The outcomes of a sequencing process in a pyrosequencer, thus consists of a read for each well, corresponding to one of the target fragments. Initially, the devices had approximately 100,000 wells, producing reads of up to 300 bp in length (i.e., 1200 cycles). Current models achieve up to one million reads, each of up to 700 bp length each.

Ion Sequencing Equipment by Thermo Fisher Scientific

Other devices that operate in a similar fashion are known as Ion based, and are currently marketed by Thermo Fisher Scientific. In these devices, clonal amplification and sequencing by synthesis are performed just like in the pyrosequencer, yet the method used to detect the incorporation of nucleotides into each of the wells differs. In this equipments, the well matrix is an extremely sensitive semiconductor that detects potential changes in each well, effectively making each one operates as a tiny pH meter. Upon evaluating the nucleotide incorporation reaction into the synthetized DNA strand, we can observe that not only a pyrophosphate is released but also a proton (H+). Considering that the volume of the well is extremely small and there are many identical DNA molecules (due to the clonal amplification), the effective concentration is substantial and the released H+ lead to a significant decrease in the well pH, which is detected by the equipment. Similarly to the previous device, a fragment is sequenced in each well, yielding a read from each well. The readings tend to be short, typically ranging from 100 and 300 bp. It should be emphasized that both "pyrosequencers" and "Ion" devices may encounter difficulties when detecting homopolymeric sequences, meaning several identical letters occurring consecutively, since it is possible that during the corresponding synthesis cycle the signal could saturate the detector, making it challenging for the device to determine whether the light intensity, and/or pH change is indicative of two, three, or more bases being incorporated.

Illumina's Platform: The Most Widely Used and Versatile High-Volume Sequencing Equipment

Let's now discuss the platform represented by the highest market volume, the most used and versatile, which is also more similar to the original Sanger method - the Illumina technology. Like in the previously described cases, the first step after obtaining the target DNA fragments is to add the adapters that allow the machine to recognize them, and perform the clonal amplification. In Illumina devices, the fragments are absorbed onto a surface that is the heart of the device: the flowcell. Here, they are arrayed much like blades of grass. This adsorption occurs because the surface has probes attached that are complementary to the adapter DNA, which then captures the target fragments through DNA hybridization. Following, clonal amplification is conducted through multiple PCR cycles. The copies of each original sequence are captured by adjacent adsorbed probes, thus creating a cluster of copies of each original target fragment. The size of the flow cell determines the number of absorbed fragments and thus the number of clonal clusters, which will correspond to the number of sequenced fragments and, consequently, the number of obtained reads.

For the sequencing itself, let's focus on one cluster. As with the Sanger method, sequencing is done by synthesis, and the incorporated base is identified by fluorescence. The key in an Illumina sequencer is that in each cycle, all four nucleotides, each with a different color, are added at the same time. This is possible, since the used nucleotides are reversible terminator nucleotides. They have a blocked 3′-OH group so when the polymerase incorporates one nucleotide into the nascent strand, it halts. As an example, let's say we are in the first cycle, the polymerase has incorporated the first nucleotide, say a "T," and since it cannot continue, it stops there. Now all molecules in the cluster should have a nucleotide T as the first newly synthetized base, while in other clusters, the corresponding nucleotide has been incorporated. How do we identify which nucleotide each cluster incorporated? - Well, as mentioned previously, the nucleotides each have a different color, so after each synthesis cycle, the fluorescence of each cluster is determined to identify which base was incorporated. Once the color (and hence the base) reading is done, a nucleotide reactivation cycle is initiated. During this cycle, an enzyme cleaves the 3′-OH block allowing the incorporation of a new nucleotide in the subsequent cycle. This reaction also cleaves the fluorescent probe from the nucleotide, so that it does not interfere with the reading of the next. In this way, in each cycle of synthesis, reading, and reactivation, a base is read in each of the clusters. The various Illumina devices differ in the size of their flow cell and thus in their sequencing capacity, ranging from the smallest to the largest: iSeq, miniSeq, MiSeq, NextSeq, HiSeq, and NovaSeq. The flow cells can sequence from about 100 million to several billion (thousand million) reads. Readings range from 100 to 150 bp for the larger devices and up to 300 bp for the smaller ones.

MGI

The latest sequencing platform of this type we will examine is marketed by MGI, a Chinese company aiming to compete in the market against Illumina by offering medium and large-scale devices. MGI sequencers also perform clonal amplification within the flow cell of the device, but the key difference is that they do not generate "clusters" of copies (like Illumina). Instead, they produce what is known as a DNA nanoball through circular PCR. In this type of PCR, the fragment, after the addition of adapters, is circularized so that the polymerase continues replicating the fragment repeatedly in the same cycle, generating tandem repeats that form a continuous strand tangled like a ball (or knot), hence the name nanoball. The result is that each "cluster" of the target fragments is a single DNA molecule, presenting the same sequence repeated over and over again. Sequencing is also performed by synthesis, but instead of using polymerase and terminator nucleotides to ensure the incorporation of only one nucleotide per cycle (as in Illumina), it is done using probes that have 5 generic nucleotides + 1 specific at the 5′ end (say a T). These probes pair with the complementary strand adjacent to the "primer" (and/or the strand being synthesized) whenever the specific nucleotide (T) can pair with the complementary strand. subsequently, a ligase joins the probe to the nascent strand, which now has one more nucleotide, in this case a T, plus 5 generic nucleotides. The probes do not possess a free 3′-OH end, so a second probe cannot join, ensuring that only one probe (with the corresponding nucleotide, the T) is added in each cycle. Similar to the case of Illumina, the four probes with different terminal nucleotides (A, C, T, or G) have a different colored fluorophore, allowing after the synthesis cycle to identify which nucleotide was added in each cluster. Once the base of all clusters is read, a reagent is added that removes the generic nucleotides (and the fluorophore) leaving the 3′-OH end free on the nascent strand for a new cycle of synthesis by ligase, reading, and removal of generic nucleotides from the probe. This type of equipment is also capable of generating millions of reads ranging from 100 to 300 bp.

Second Generation Sequencing Key Points

The four platforms we have described are recognized in contemporary science as second-generation sequencing techniques. They currently dominate the market, particularly in the field of Clinical Genomics. As we noted earlier, they share several common features which include: (i) the requirement of clonal amplification of the target fragments; (ii) employing sequencing by synthesis; (iii) they sequence through cycles of reaction(s) combined with detection; (iv) sequencing multiple fragments simultaneously in very small volumes; and (v) producing relatively short sequences compared to the traditional Sanger method. Understanding the reasons and implications of these five features is critical, as developing alternatives to them is key to advancing third, and possibly fourth-generation sequencing technologies. Let's examine these aspects one by one.

Clonal amplification aims to increase the copy number of each target fragment until a threshold is met where the corresponding signal (light, pH, or fluorescence) can be effectively detected. In devices like those made by Illumina, each cluster holds thousands of molecules, and synthesizing them represents a significant expenditure on reagents. Moreover, clonal amplification can introduce copying errors, which might result in a read that does not accurately reflect the original sequence. Specifically, in Illumina devices, an issue of fragment duplication can occur. This happens when a copy drifts and forms a new cluster at a distance, leading, ultimately, to two readings that actually correspond to the same fragment. This is known as a duplicate read. Sequencing by synthesis also heavily influences the cost of reagents, which encompasses both nucleotides and the enzymes used in the synthesis process, as well as activation and/or detection (as seen in pyrosequencing). The fact that sequencing occurs one nucleotide at a time makes these devices relatively slow. Sequencing a batch of 100–300 bp can take several hours, whereas a polymerase alone could synthesize a strand of that length in mere seconds. The extensive use of parallelization and microfluidics grants second-generation sequencers their remarkable efficiency, while the reduced reaction volumes mean less reagent use and, consequently, lower costs.

Lastly, it's vital to grasp why these machines generate short reads. Note that in all these devices, sequencing by synthesis does not occur in a single molecule but in a cluster containing thousands of copies. Considering Illumina's technology as an example. Imagine that in a specific cluster, an "A" corresponding to the color "red" is added in the initial cycle. The reader detects red fluorescence and assigns "A" as the read's starting letter. If the cluster includes 1000 molecules with 99% synthesis efficiency, 990 will have incorporated an "A" after the first cycle, while 10 cluster molecules lag behind. In the subsequent cycle meant for a "C", 980 might incorporate "C" (blue), with the out-of-sync ten adding an "A" (red). Although the detector observes mostly blue at this stage, as cycles progress and synchronization within the cluster diminishes, each cycle adds a nucleotide mix, leading inevitably to a color blend. By the 100th cycle, roughly 900 molecules might add the correct nucleotide, like "T" (yellow), but the remaining 100 add a blend of A (red), T (blue), and G (green). At this juncture, the read's quality deteriorates below acceptable thresholds due to uncertain color observations, prompting the systems to halt beyond 100–300 base pairs. This synchronization issue is why reads from these devices are comparatively short, and thus are called short-reads.

A summary of the paradigm shift of massive sequencing technologies rooted in the simultaneous miniaturization of reaction volumes and the massive parallelization of the sequencing-by-synthesis process, is illustrated in Fig. 3.1. Advancing these aspects furthers the evolution of sequencing methods, and significant changes or eliminations of these features often herald the emergence of a "new" generation of sequencers. Before we transition to discussing third-generation sequencers, let's delve deeper into the concepts of reads, and another important concept the so-called "pair-end."

The Concept of Read and The Concept of "Pair Ends"

As we have already discussed, the read is the fragment of the target sequence that is read by the sequencer during each reaction. Let's explore how it is related to the target fragments, which we previously explained derive from the sample and contain the

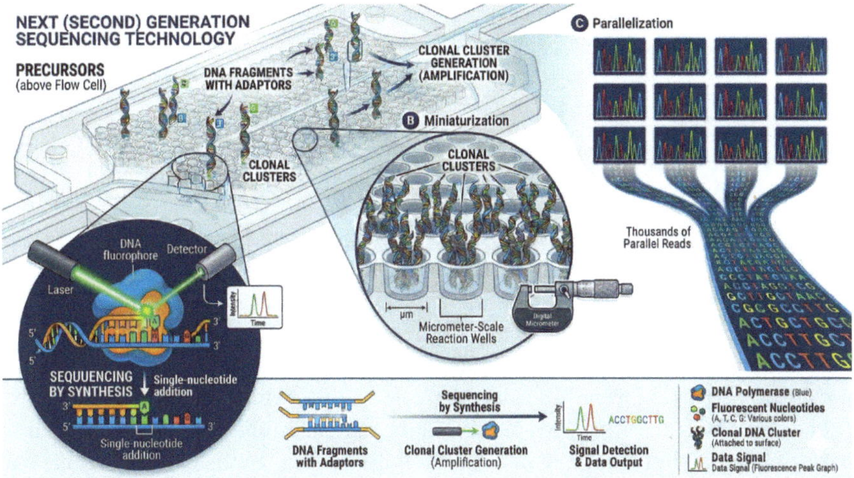

Fig. 3.1 Principles of next (Second) generation sequencing (NGS). The transition from raw DNA to digital data involves several key stages: (**A**) Precursors: DNA fragmentation and the ligation of specific adaptors. (**B**) Miniaturization and Clonal Amplification: DNA fragments are captured on a solid surface (flow cell). (**C**) Sequencing by Synthesis

desired target sequence. In a device equipped for second-generation sequencing, the target fragment is double-stranded DNA with adapter sequences at both ends. During sequencing by synthesis, one of these strands will serve as the template, the primer will bind/anneal to the adapter, and therefore, the read encompasses the 100–300 bases starting from where the adapter ends (the original end of the target fragment) to the total number of bases in the corresponding read.

Example Figure: Primer: *5′-GGTACAGT* ACGACAGTACAGT-3′ read being synthetized AdaptER:*3′-CCATGTCA* TGCTGTCATGTCACAAA TGTTGCCCATGGGTCAAACTTTGGTTGAT-5′ Template strand from fragment

The same processing occurs on the other strand of the target fragment, initiated from the opposite end. This creates a second read, corresponding to the sequence at the other end of the target fragment. Thus, for each original target fragment, two readings are generated that relate to the sequences at both ends.

Example Figure: Template strand of the Target Fragment (complementary) ACAGTACAGTGTTTACAACGGGTACCCAGTT*TGAAACCAACTA-3′* Adapter read being synthetized 3′-TGCCCATGGGTCAA*ACTTTGGTTGAT-5′* Primer

It is crucial to understand that these two sequence reads are not independent because they both stem from, and relate to, the same target fragment. It is commonly stated in genomics terminology that the readings are paired, or are pairs. As their sequences represent the ends of the target fragment, each set of these readings is known as a "pair-end." In second-generation sequencing devices, readings typically are sequenced and gathered in this manner, resulting in two sets of paired reads (each paired with its mate). Information from the pair-ends is fundamental for sequencing assembly and proves exceedingly valuable in numerous applications.

Third Generation Sequencing

As Next-Generation Sequencers evolved into the current (second) generation, the question arose: what defines a third-generation sequencer? Transitioning across generations implies a significant alteration in the DNA reading process. From the points previously discussed, it is evident that the abandonment of clonal amplification and/or sequencing by synthesis were the leading options, which could potentially result in reduced costs, heightened speed, and greater read lengths.

Pacific Biosciences (PacBio) pioneered the march in this regard by wholly eliminating clonal amplification, which substantially reduces both cost on reagents and processing time. However, this approach necessitates the ability to detect the signal that enables the reading of a base at the single-molecule level. Techniques for single molecule detection and tracking, which have been honed over the past two to three decades, generally depend on powerful fluorescence microscopes. Owing to the exorbitant costs of integrating these microscopes into a benchtop sequencer, the developers at PacBio adopted an alternative strategy borrowed from materials science to amplify the signal. In PacBio systems, the sequencing wells contain a nanostructure designed to enhance the fluorescence signal through a phenomenon known as "Zero-Mode-Waveguide (ZMW)." ZMWs, cylindrical or nanophotonic structures, are crafted by creating extremely tiny holes in a metallic layer deposited onto a substrate (like glass or silicon). The smallness of these holes (diameters in the tens of nanometers) prevents support for light propagation modes in the optical wavelength range, resulting in "zero-mode" confinement of light that allows for precise detection and analysis of single molecules, drastically lowering background illumination and improving the signal-to-noise ratio, even amidst high fluorophore concentrations in the sequencing environment.

Within the chambers of PacBio machines, a polymerase molecule, is fixed at the well bottom, and the target fragments with generic primers and fluorescent nucleotides are added to the flow cell (i.e., the reaction chamber). Specific nucleotides ensure that the fluorophore detaches when the polymerase integrates it into the synthesizing strand, which happens since the fluorophore is linked to the gamma phosphate of the triphosphorylated nucleotide. Each integration of a nucleotide into the well, and its subsequent incorporation by the polymerase, leads therefore to the release of the fluorescent molecule, and emission of the fluorophore, with the signal captured by the device in real-time.

This sequencing process, also synthesis-based, occurs in real-time (thus the SMRT acronym, for Single Molecule Real-Time sequencing), allowing for rapid sequencing of hundreds of bases. Moreover, as it sequences a single molecule (rather than a cluster), there are no alignment issues, and the reads are notably longer. They extend as far as the target fragments or until the polymerase disengages from the DNA template being replicated. This results in read lengths ranging from thousands of bases (kilobases, kb) to millions (megabases, Mb).

Therefore, considering the significant advantages of PacBio systems — notably their capacity for longer reads and faster processing — versus second-generation machines, their shortcomings merit further exploration. The primary disadvantage is the error rate. Whereas second-generation machines exhibit a typical error rate of between 0.1% and 1%, PacBio machines often range from 1% to 5%. This higher rate stems from the rapidity and reliance on detecting single-molecule fluorescence, complicating accurate base identification. To mitigate this, PacBio machines may circularize the target fragments, sequencing them several times to reduce errors, albeit at the cost of increased reagent consumption with each duplicated sequence. It is also crucial to note that PacBio machines are two to five times more expensive than second-generation devices, posing a significant barrier to widespread adoption and acquisition.

Is Direct DNA Reading Possible?

The last method we will examine in some detail, also considered a third generation method, is developed by Oxford Nanopore, which entails sequencing DNA without performing clonal amplification—that is, using single molecule techniques—but also without synthesizing a new strand. Eliminating synthesis sequencing virtually erases the total cost of reagents (nucleotides and polymerase), and often a method capable of achieving this task, has been deemed the "holy grail" of sequencing.

The crux of the issue in direct reading (i.e., without synthesis) is how each of the DNA strand bases is identified and distinguished. Traditionally, these ideas were based on either degrading the DNA one base at a time (similar to how proteins are sequenced), or somehow having a device that moves along the DNA strand, determining which letter it is seeing (i.e., reading) as it moves. This method is analogous to a "scanner," or how a person reads a text, that is, following the sequence of characters (letters) in order, and identifying them. Thus, to directly read DNA, I must be able to: (1) Look at the bases one by one (i.e., in order), and (2) identify them. Among the various techniques used to scan DNA, the most effective and successful has turned out to be the use of nanopores. Let's see how they work.

DNA is a highly charged molecule, so when it is subjected to an electric field, the molecules are forced to move according to the orientation of the field. In fact, this is the principle of electrophoresis that we mentioned earlier in the context of Sanger sequencing. As a charged molecule, DNA cannot cross a lipid bilayer. Therefore, if we place DNA at the top of a compartment separated from the bottom by a lipid bilayer, the DNA will be retained in the upper compartment. If an electric field is applied that forces the DNA to migrate to the lower compartment, it will be blocked by the bilayer. But if a "pore" (a hole) is introduced in it, the DNA will accommodate itself to pass through the pore, thus reaching the lower compartment. The phenomenon is similar to that occurring in our neurons, where a voltage difference (i.e., an electric field) drives ions, which enter/exit the neurons when voltage-gated ion channels open, propagating the action potential and transmitting the nerve impulse.

The migration of DNA through pores occurs for both single and double-stranded DNA and has been known for several decades. The pores used are protein pores and their nanoscopic size earns them the name "nanopores." When DNA passes through the pore, it does so "stretched out," so if one were to "look," for example, at the center of the pore, one would see the bases (or base pairs if it's double-stranded DNA) pass one by one, and in order. This solves the first problem for sequencing, but now what is needed is some method that can identify, one by one, which bases are passing through the pore.

Various developments have been attempted to identify the base passing through the pore. These methods include: the removal of the base and subsequent identification by mass spectrometry techniques, and identification by spectroscopic or optical techniques (remember that it is a single molecule, so the signal intensity problem, as in PacBio devices, is present). However, the successful method does not rely on any of these but uses an indirect measurement, again borrowed from neurobiology. It's important to note that when DNA passes through the pore, it is accompanied by positive ions (mainly $Na+$), which are also driven by the electric field. Like in neurons, the flow of $Na+$ ions through the pore generates an electrical current that can be measured. In the total absence of DNA, the ions subjected to the field pass through the pore unimpeded, and this maximum flow thus generates a maximum, stable, and continuous current.

When a DNA molecule starts to pass through the pore, it partially blocks the ion flow, which is reflected in a drop in the observed current. Taking these observations as a starting point, researchers at Oxford Nanopore decided to design a pore that would allow single-strand DNA to pass (in addition to $Na+$ ions) and that would interact differentially (in terms of strength or affinity) with each of the four DNA bases (A, C, T, or G). The successful design results in a pore that is differentially blocked by each of the four bases. The base that binds the strongest, in this case "G," is the one that blocks the pore for the longest time (producing the largest drop in the observed current), while "C," which binds weakly, blocks it much less, resulting in a relatively minor decrease in the current that is registered.

When this pore is used under an electric field with DNA molecules and $Na+$ ions, the ion flow generates a continuous current, but when DNA begins to migrate through the pore and a base "fits" into the "neck" of the pore (the "neck" being the narrowest section of the nanopore), it blocks the pore, and a drop in the current is observed. When the base is released, the current returns to its original value, until the next base blocks the pore's neck. In this way, the DNA bases block the pore one by one. Since each one blocks the pore with different strength, the drop in current varies for each of the four bases, allowing them to be identified and thus sequence the target fragment.

Like the devices at PacBio, in Oxford devices, the readings are long, as they sequence a single molecule. Oxford devices also have less accuracy than second-generation devices since the signal (current drop) of the different bases has some overlap. But what's most interesting about Oxford devices is that they do not require reagents for synthesis or clonal amplification, and that the devices are extremely cheap and small. A small device, the minION, is the size of a cell phone and operates connected to a PC via USB. The main drawback of Oxford devices is that as the pores are proteins, they are unstable and after a few rounds of sequencing the pores stop working. Typically, the flow cell of an Oxford has about 1000 pores, of which it is estimated that between 50% and 70% function optimally, as another problem with working with protein pores is their variability, due to the complex system for their synthesis. In this sense, the flow cell and the pores are considered "disposable" and represent the main cost of sequencing, so they are often regarded as the cost of reagents. Currently, the cost to sequence, let's say 1 Mb of DNA with a nanopore due to the cost of the pore, is similar to the cost of reagents in second-generation devices. The fact that no initial hardware purchase is required makes Oxford nanopores an excellent alternative to entering the world of massive sequencing. However, it is important to highlight that they are currently used mainly in applications associated with microorganisms, and to a much lesser extent in applications of clinical genomics in humans.

The Future of DNA Sequencers

Although, as we already mentioned in Chap. 1, predicting the future is no easy task. An analysis of recent advances and publications on the subject (including patents), combined with knowledge of the main challenges presented by current sequencing technologies, allows us to venture two axes of future DNA sequencing developments. The first involves the incremental advancement of second and third generation MSTs. Although not resulting in a paradigm shift, the optimization of clonal amplification processes, synthesis sequencing, and detection is expected to continue expanding capabilities and reducing costs. In this area, robotics and automation of processes will play an increasingly major role, and the reduction in the use of reagents through greater miniaturization, or reduction of discard between cycles will contribute as well. On the other hand, third-generation technologies are expected to continue improving their precision. However, none of these advances seem to seek a paradigm shift that would lead to fourth generation technologies. In that context arises the following question: What could represent this change?

The holy grail of sequencing technologies is one that allows direct and accurate reading of long DNA fragments, without the need for synthesis and reagents, and in real time. Nanopores are the technology that most closely resembles this ideal, but they still have a clear weakness: the pore is protein-based. The use of proteins (polymerase, ligase or the pore itself) is a common point among all existing sequencing methods, and eliminating this, giving way to a non-biological sequencing methods, would possibly be considered the next paradigm shift. There are two approaches geared towards these developments. On one hand, the fixing of target fragments onto a solid support that would allow them to be "looked at." and by optical or spectroscopic methods determine the ordered bases. It would be somewhat analogous to determining the sequence of the DNA fragment by X-ray crystallography, in a hyperextension of what was done by Watson and Crick. The current problem with this approach is that there is still no determination technique (X-ray, cryoEM, ultra-high resolution microscopy) that can efficiently determine the sequence of millions of fragments (quickly, and at low cost), and that can be available on a desktop device. Today the mentioned techniques require larger-scale equipment.

The other approach, perhaps more promising, is based on replacing the protein nanopore with one made of nanomaterials, such as those used in electronic components or microchips. This electronic nanopore could allow, just like its protein counterpart, the linear passage of single-strand DNA, which coupled with a technique that allows identification of the passing bases, would result in a non-biological sequencer, that could potentially be more accurate and durable than the current nanopores. This would undoubtedly result, not only in a paradigm shift but also in a significant further reduction of costs, potentially bringing the cost of sequencing itself per se to zero. Having described the operation of the main sequencing technology equipment, we will now move on to describe the steps associated with a standard sequencing experiment and its alternatives.

Outline of an "Experiment" in Massive Sequencing Technology

The starting point for an experiment using massive sequencing technology, often referred to as a "run" (from the direct translation of the English term "run" when the sequencer is in operation), is always the purification (also called extraction) of DNA. This step is crucial and can condition the quality of the run's result since if the DNA degrades or part of it is lost during the process, this will be reflected in the results.

Once the DNA has been extracted, the second step consists of what is called library preparation. The library is the set of target fragments, with their adapters, that will be introduced into the sequencer for reading. In other words, preparing the library involves determining which target fragments will be used and what characteristics they will have. It is here where experimental design makes its principal contribution and where a world of possibilities opens up that give rise to the different techniques or applications of massive sequencing. In an experiment where the aim is to sequence all the DNA in the sample uniformly, preparing the library simply involves fragmenting the DNA to an appropriate fragment size and adding the adapters. This "simple" version is used for sequencing a complete genome (Whole Genome Sequencing, WGS), whether it is that of an individual, a bacteria, or another organism of interest. In this case, the experimental design consists only of purifying the DNA to contain only the desired genome and fragmenting it according to the optimal size for the sequencer and the desired run's objective. The variation in fragment size is something that will be discussed later, as it has specific applications.

The alternative to sequencing all the DNA contained in the sample involves carrying out a "selection" of it according to some particular interest. There exist two main techniques for this process, known as: (i) Amplification, and (ii) Selective Capture, respectively. Selective amplification involves carrying out a PCR that selectively amplifies one (or more) region(s) of interest. The paradigmatic case of using selective amplification in clinical genomics is known as sequencing a "single gene," or "a panel" of a few genes. A panel is a small collection of 2–50 genes of interest. In this case, primers that can anneal at the same annealing temperature must be selected for each of the target sequences. For instance, if all "5" exons of a gene are being targeted, pairs of primers are used one for each exon (provided the exons are not too long). This PCR, which amplifies numerous fragments, is known as multiplex PCR. The problem with multiplex PCR is that the number of different fragments it can amplify from the same sample (raise) has a limit, in the order of tens, as afterward amplification starts to become uneven. That is why multiplex PCR amplification is used for relatively small gene panels or for applications where the same fragment is lifted from many samples (we will see this a bit later).

In the case of selective capture, the original DNA is not amplified, but is first fragmented and then mixed with a set of probes that define the fragments to be captured. These probes have some type of device that allows them to be separated from the mixture. One of the most used techniques to separate the probes with the captured fragment from the rest of the fragments involves coupling the probes to a metallic sphere (magnetic beads). Then using a powerful magnet, below the container of the fragment plus probes mixture, these will be dragged to the "pellet," leaving the fragments to be discarded in the supernatant. Affinity columns, among other methodologies, are also often used. Each probe consists of single-stranded DNA, usually 10–20 bp, which is complementary to the corresponding bases of the region of interest. For example, if we want to capture exon 1 of a given gene, we would design a probe with complementary sequences to a 10–20 bp region of the corresponding exon. When the DNA sample is mixed with the probes, these hybridize to all the fragments that have the DNA complementary to them, i.e., the DNA of interest. It is important to note that while the probe, although measuring 10–20 bp, will bind to a fragment that may be from one to tens of kb, thus capturing the region of interest and its context. If one wishes to capture, say, all the exons of a gene, then a collection of probes must be designed so that there is one complementary to each of the desired exons. The advantage of capture over multiplex PCR is that one can use hundreds or thousands of probes since the hybridization is done slowly and at low temperatures. This technique allows, for example, designing a collection of probes that are complementary to about 100,000 (or more) exons of the human genome, with which the capture will obtain the whole exome. If this set of captured sequences is sent to the sequencer, it is what is known as Whole Exome Sequencing (WES). The WES is possibly the most used sequencing technique in clinical genomics. Regardless of whether only fragmentation, amplification, or selective capture has been performed, what is obtained after the library preparation process is a set of target fragments that have adapters coupled at both ends that are required to be sequenced as seen in the previous section. There is an additional point of complexity associated with experimental design, which allows, in the same sequencing run, to sequence different samples. To read different samples in the same run, what must be done is to prepare the libraries of each sample separately until the fragments are obtained. At this moment, when the adapters are coupled, a sequence of about 10 bp called a "tag" is also added to them, which will be different for each sample. In this way, each sample will have a part of the adapters that is the same, the one required by the sequencer to identify it as a target fragment, and a region "the tag" which is specific to each sample. Then, the samples are pooled together and introduced into the sequencer all together for reading. As they are read, the samples are identified individually according to the reading of the tag region. The "tagging" of the samples is what allows a large-scale device to sequence in the same run, for example, tens of exomes, or from hundreds to thousands of panels in the same run. With this section, we conclude the description of the technical aspects of massive sequencing technologies, and we should then move on to the processing and then analysis of the data or "Bioinformatics," which will be seen in the next Chapter. However, in the next section, we will leave some reflections related to the evidence of the paradigm shift produced by massive sequencing techniques, and their impact both on the scientific community and on society in general.

Where is the Paradigm Shift of MSTs Evidenced

The major evidence of a paradigm shift brought about by massive sequencing technologies, which were initially called next-generation technologies, can be observed in a sequencing cost chart (in logarithmic scale) over time, for the last 25–30 years. When analyzing the capacity and/or cost of most technologies in a log scale over time, in the vast majority of cases, a linear trend is observed. This observation when analazing the number of transistors in a microprocessor, which doubles every 2 years, led Moore to propose his famous law. Moore's law generally expresses that the growth of a technology's capacity follows an exponential growth, resulting in a linear graph on a logarithmic scale. Consequently, costs for the user usually reduce proportionally. If we analyze the corresponding graph for sequencing costs, we observe that between 2001, when the first human genome was sequenced using Sanger technology, until 2007 when massive sequencing techniques were introduced, the fall in cost on a logarithmic scale is very close to linear. This provides evidence of the improvements, and incremental cost reductions associated with Sanger sequencing. However, from there, in a period of 5 years, the cost fell dramatically, by about a thousand times. This impressive drop in cost over such a short period evidentially represents that next-generation sequencing is a paradigm shift. Since 2012, the cost has again remained stable, assuming a linear trend but with a slope slightly less than the initial one, which describes the current incremental advances in second and third generation massive sequencing technologies.

To appreciate the magnitude of the drop over a period of 5 years, of three orders of magnitude, we can use the following analogy. Undoubtedly, if you are interested in cars, you would have noticed that over time for say around 10,000 USD (the cost of an average car) the vehicle one can acquire is increasingly better. Additionally, if you also like cars, you surely have dreamed of owning a Ferrari, whose current cost we can assume to be close to one million dollars. Imagine, then, that in 5 years you could acquire a Ferrari (or a similar capacity car) for less than a thousand dollars. What do you think? Wouldn't that be a paradigm shift in the automotive industry?

To continue appreciating the magnitude of the drop, let's analyze some numbers. If we take as a reference cost the first human genome sequenced privately (the one sequenced by Celera Genomics) this cost, in the years prior to 2001, was about 100 million USD. Six years later in 2007 the first genome sequenced by massive sequencing technology would cost one million dollars, and the following year it was offered at less than 100 thousand. By 2010 a complete human genome costed less than 10,000 USD, and by 2014 it surpassed the barrier of 1000 dollars. Recently, towards the end of 2023, a new company announced that its sequencers would allow sequencing a human genome for about 100 dollars. With those values, aren't you tempted to sequence your own genome?

The Future Impact of Massive Sequencing Technologies on Society

To understand the potential future impact of massive sequencing technologies, it is interesting to compare them with two technologies that have undergone innovative paradigm shifts in their costs and capacity. One of these is the personal computer (PC), and the other is the cellphone/smartphone. One of the clearest trends observed in PC and smartphone technologies is a continuous increase in their acquisition by society in general (who does not have a PC at home and/or a smartphone these days?). In genomics, while the progress has not been as rapid as that of (i)phones, there has been a noticeable increase in the number of sequenced genomes over the last 5–10 years, as well as an acceleration in the adoption of the technology. This implies, unequivocally, the following: No matter how useful, ethical, affordable, accessible, etc., sequencing an individual's genome is, in the future, there will be more and more individuals with their genomes sequenced. In fact, it is estimated that currently between 1 and 5 million genomes have been sequenced, which approaches 0.1% of the global population. In this context, and again regardless of what our "sentimental," "political," or "ethical" position on the usefulness of sequencing human genomes may be, it is clear that there will be a growing need (or demand) in the near future, for those professionals capable of interpreting personal genomic information, especially in its link with our health, clearly meaning a greater demand for professionals in clinical genomics.

However, in my personal opinion, the most interesting aspect when comparing genomics with PC and/or smartphone technologies lies in the analysis of the current uses of these devices compared to the purposes for which they were originally conceived. The original PC emerged from the need of an IBM technician to continue working at home (What a need!). Previously, computers were only available in centralized offices of large institutions. Although initially it played a role primarily associated with work, which it still maintains today, it is clear that PCs have also become a multimedia entertainment center, a communication tool (both social and work-related), among many other uses. Similarly, the cellphone (or portable phone) was conceived as a tool for remote communication. But by equipping it with internet, the explosion of so-called "apps" turned it into a multipurpose device. Both in PCs and in cellphones, it is unlikely that at the time of their conception the developers could have imagined the world of possibilities that would open up with their mass adoption. This "mutability" and expansion of utilities is another characteristic of certain products that become massive. Returning to the human genomes then, the reflection I propose is the following: Although the human genome has been considered from its beginnings as a tool to help improve our health, what new applications or uses can we imagine will be found for it in the future? Is it possible to imagine a future where the use of the genome in health is not the main one?

With this reflection, we conclude our chapter on the techniques, technologies, and strategies for reading DNA and prepare to venture into the world of bioinformatics in the next chapter.

References and Suggested Further Readings

Sanger F, Nicklen S, Coulson AR. DNA sequencing with chain-terminating inhibitors. Proc Natl Acad Sci USA. 1977;74(12):5463–7. https://doi.org/10.1073/pnas.74.12.5463.

Margulies M, et al Genome sequencing in microfabricated high-density picolitre reactors. Nature. 2005;437:376–80. https://doi.org/10.1038/nature03959.

Bentley DR, et al Accurate whole human genome sequencing using reversible terminator chemistry. Nature. 2008;456:53–59. https://doi.org/10.1038/nature07517.

Eid J, et al Real-time DNA sequencing from single polymerase molecules. Science. 2009;323: 133–8. https://doi.org/10.1126/science.1162986.

Clarke J, et al Continuous base identification for single-molecule nanopore DNA sequencing. Nat Nanotechnol. 2009;4:265–70. https://doi.org/10.1038/nnano.2009.12.

Heather JM, Chain B. The sequence of sequencers: the history of sequencing DNA. Genomics. 2016;107:1–8. https://doi.org/10.1016/j.ygeno.2015.11.003.

Shendure J, et al DNA sequencing at 40: past, present and future. Nature. 2017;550:345–53. https://doi.org/10.1038/nature24286.

Nakano M, Nakamura K. Next-generation sequencing: current technologies and applications. CRC Press; 2023.

Chapter 4
Bioinformatics

Abstract This chapter introduces Bioinformatics as a fundamental discipline underpinning modern genomics and Clinical Genomics. It focuses on the computational methods required to transform raw sequencing data into biologically meaningful information. Starting from the output of sequencing technologies—millions of short DNA reads—the chapter explains how bioinformatics integrates algorithms, data analysis, and database organization to interpret genomic data. Historical developments are outlined through key resources such as GenBank, Swiss-Prot, RefSeq, and UniProt, emphasizing the importance of standardized biological databases. The chapter illustrates how bioinformatics works through a conceptual algorithm for identifying coding regions, introducing essential concepts such as the genetic code, reading frames, reverse complement, and open reading frames (ORFs). It then presents the two central computational challenges in genomics: de novo genome assembly and read mapping against a reference genome for variant identification. Using intuitive analogies, the complexity of reconstructing genomes from fragmented reads is explored. Finally, key quantitative metrics such as read depth and coverage are defined, highlighting their role in assessing sequencing quality and reliability. Overall, the chapter establishes bioinformatics as a critical bridge between sequencing technologies and clinical interpretation, enabling the extraction of actionable knowledge from genomic data.

Keywords Bioinformatics · Genome assembly · Read mapping · Open reading frames · Coverag

What About Bioinformatics?

Prior to advancing in our journey towards understanding Clinical Genomics, we must divert slightly and make a stop at the second discipline that ensures the success of genomics: Bioinformatics. As we recall from Chap. 2, the result of a

M. A. Marti, *Clinical Genomics*,
https://doi.org/10.1007/978-3-032-34667-4_4

sequencing experiment is a collection of reads, or fragments of DNA sequences. In the case of Massive Sequencing Technologies, this involves hundreds of millions of fragments. Bioinformatics is the discipline that develops and implements techniques to process these reads, interpret their information, and convert it into biological knowledge. Basically, as we mentioned at the beginning of the book, bioinformatics translates the information written in the language of biology into plain English.

From a general point of view, Bioinformatics is defined as the discipline that develops and utilizes computer techniques in a biological context. In other words, it consists of the development and application of algorithms and computer programs working with biological data. In practice, it is organized into three levels: (i) the development per se of theories and algorithms, (ii) the use of these algorithms for data analysis, and (iii) the organization and classification of biological data. Examples of these levels in the realm of Clinical Genomics would be: (i) the development of algorithms to determine variants from the reads derived from an individual's genome, (ii) the use of programs to understand and predict the effects of relevant variants, (iii) the organization of the variants in a database like ClinVar or gnomAD.

Bioinformatics began its early development around the year 1960 when Margaret Dayhoff and her collaborators compiled all the known protein sequences of that time, along with information related to their functions and biological context (known as annotation), and created the Protein Information Resource. This resource later led to the "Atlas of Protein Sequence and Structure" which was distributed in print form during the 1960s and 1970s. In the early 1980s, Amos Bairoch, one of the fathers of bioinformatics, created the Swiss-Prot database within the European Molecular Biology Laboratory (EMBL), incorporating the information from the Atlas. In 1986, Bairoch distributed Swiss-Prot, which at that time contained information on approximately 4000 protein sequences, through the precursor of the internet, marking a foundational milestone for the discipline.

In parallel to the development of Swiss-Prot, the EMBL also established a database to store nucleotide sequences, primarily DNA derived sequences from first-generation (Sanger) sequencing experiments. Concurrently, the National Center for Biotechnology Information (NCBI) in the United States created GenBank, and the Japanese established the DNA Database of Japan (DDBJ), all three sharing with the same objectives. These three databases began to collect and catalog DNA sequences and share them openly. Finally, in 1988 they convened to create the International Nucleotide Sequence Database Collaboration (INSD), a consortium defining the basis and the format for storing nucleotide sequences. Among other things, they are the creators of the quite famous "fasta" format.

The accumulation of sequence information, its redundancy, and its lack of precision, led the NCBI in 2003 to begin the RefSeq project, aiming to define the "reference" sequence for each biological molecule. Today, RefSeq contains the reference sequences of genes (and genomes), mRNAs, and the resulting proteins. In 2004, another significant bioinformatics initiative resulted in the development of UniProt, a database created to unify all available information regarding protein sequences, and now serving as the reference database for proteins, and their functions.

It's important to mention that Bioinformatics starts, but does not end, with sequences. A second branch of bioinformatics studies the shape of biomolecules, the three-dimensional structure in space of DNA, RNA, and primarily proteins, in addition to a myriad of metabolites and various molecules that constitute life. This field also traces its origins to the determination of the DNA structure by Watson and Crick, although it was the subsequent determinations of the protein structures of Myoglobin and Hemoglobin by Kendrew and Perutz in the 1970s that marked the birth of structural biology. The complexity of biomolecule structures is insurmountable without the use of bioinformatics tools that allow us analyze, visualize, and understand them. In this context, it is perhaps important to emphasize that one of the most significant developments in biological sciences in the recent years, belongs to this domain: the development by a Google (Alphabet) of an "Artificial Intelligence" algorithm capable of predicting the three-dimensional structure of a protein from its sequence. This development, called AlphaFold, represents a major revolution in structural bioinformatics and, as we will see later, also has an impact on Clinical Genomics, particularly in relation to the process of interpreting missense variants.

Finally, I wish to discuss perhaps one of the most advanced areas of bioinformatics today. This relates to the algorithms that seek to understand the complexity of the specific metabolic state of a cell (or a set of them, for example a tissue) based on data related to its expression profile (obtained either by microarrays or RNA sequencing), its set of proteins (proteomics), and/or its metabolites (metabolomics). This area of bioinformatics, characterized by its distinctive approach, seeks to understand biomolecules not individually, or in small interacting groups, but as a whole in a holistic manner. In other words, they describe the cell as a network of entities (genes, proteins, metabolites) that interact dynamically and ultimately define the cellular state and beyond.

"How Does Bioinformatics Work?"

To gain an initial understanding of how bioinformatics works and to approach the concept of an algorithm in a bioinformatics context, let's consider one of the principal bioinformatics algorithms together. This algorithm aims to accomplish the following: "Given a DNA sequence, the algorithm should be able to predict whether any region of it (in other words some part of the sequence) might correspond to a coding region, thus encodes for a protein (or a part of it)."

To begin conceptualizing our algorithm, we first need to frame it in a proper context. This involves understanding the relationship between DNA sequences and the proteins encoded by them. This brings us to the context of what is known as the "central dogma" of molecular biology. According to this dogma, DNA is "transcribed" to RNA, and then this is "translated" through the genetic code into the protein sequence. An analysis of both processes shows that while the DNA => RNA step is straightforward, merely requiring only the replacement of Thymine (T) with Uracil (U) without altering the sequence's information content, or needing external knowledge. The second step requires a dictionary to "translate" the RNA sequence into a protein sequence. This dictionary is the genetic code, which establishes correspondences between "triplets" (often referred to as codons) of RNA or DNA, sequences of three bases (letters), and the resulting amino acid. This code is readily available on the internet, but let's show here a small extract of it (omitting the RNA step):

TTT Phe(F) TTC Phe(F) TTA Leu(L) TTG Leu(L) ... . TGC Cys(C) TGA STOP TGG Trp(W)

Using this section of the genetic code, it is easy to see how our algorithm, starting from a DNA sequence, for example: TTCTGCTGGTTATGA, could predict that, if the sequence were actually transcribed and subsequently translated, the resulting amino acid and thus protein sequence would be: Phe-Cys-Trp-Leu-Stop (you can verify this easily for yourself!).

Before we proceed, let's remember (and analyze, if it's the first time you are looking at it closely) the genetic code. There are 20 different amino acids, plus an additional signal referred to as the "STOP codon." The STOP codons, corresponding to the triplets TAA, TAG, and TGA, indicate that the translation should end, meaning no more amino acids should be added to the protein chain. Therefore, STOP codons mark the end of the protein. Considering that there are four possible DNA bases (A, C, T, and G), there are 64 = 4^3 possible different codons, which must code for the 20 amino acids + the stop signal (or stop codons). Therefore, for each amino acid plus the STOP codons, there is (sometimes) more than one triplet that can encode them. Due to this property, the non one-to-one correspondence between codons and amino acids, the genetic code is said to be degenerate. Degenerate or not, it's easy to see how our algorithm should be capable of translating any DNA sequence into the corresponding "putative" protein sequence, right?

A second crucial component of the "translation" process involves something known as the reading frame. The reading frame specifies from which letter of the DNA (or RNA) sequence, the three letter codons should begin to be assigned. To illustrate this concept, consider the following DNA sequence (DNA sequences are typically written in lowercase, reserving uppercase for protein sequences):

DNA Sequence: actggtca

One possible triplet assignment could be: act-ggt-ca, which would result in the following amino acid sequence "TG" (Threonine-Glycine). In this example, the triplets are assigned starting from the first base of the DNA, constituting the "first" reading frame.

However, if we shift by one base, we can form the following triplets (leaving the first base alone): a-ctg-gtc-a, corresponding to the amino acids "LV" (Leucine-Valine) in what would be the translation of the second reading frame.

Finally, if we start at the third base of the DNA, the resulting triplets would be (leaving the first two bases "ac" unassigned) ac-tgg-tca, corresponding in this third reading frame to the amino acid sequence "WS" (Tryptophan-Serine). If we shifted one more base, it would simply be the first reading frame starting at the second codon.

In summary, given a DNA sequence, it is possible to obtain three different protein sequences, each corresponding to one of the three possible reading frames. Thus, our algorithm, given the starting DNA sequence, first generates not one possible protein sequence, but three (one for each reading frame). These three protein sequences then need to be analyzed to determine if any region corresponds to a potential protein (coding region) or not. So what additional information can we use to distinguish potential coding regions from non-coding DNA sequences?

However, before we move into how to distinguish coding from non-coding sequences, once translated, I have to bring another issue that needs to be considered. Recall that biologically, within a cell, DNA is found as a double strand formed by the so-called "two" complementary strands. When we work with DNA sequences, we usually write (and read) only one of them, and by convention, it is written from the 5′ (phosphate end) to the 3′ (OH end of the sugar used by the polymerase to add nucleotides when synthesizing DNA). But, why can we write just one strand and "ignore" the other? Given a strand (or chain) of DNA, the other is obtained directly and unequivocally by base complementarity. Remember that pairing occurs between A:T and C:G, therefore from the simple DNA strand sequence: actgtcagt, we can easily obtain the complementary strand and write the double-strand DNA sequence this way: 5′-actgtcagt-3′ 3′-tgacagtca-5′ Notice, when written as a double strand, the complementary strand appears "reversed," in the 3′ to 5′ orientation. To write it according to convention, it would appear as follows: 5′-actgacagt-3′

Therefore, given a DNA sequence, we can use base complementarity, and the knowledge that the complementary strands run in opposite directions, to tell our algorithm to do the following: take the DNA sequence, then replace A with T, T with A, C with G, and G with C and finally reverse its order. In this way, we have now obtained the other strand, which is termed the "reverse complement." The question, you should now ask yourself is: Why is this important?

Considering the reverse complementary DNA sequence is crucial since, in a biological system, when we read a DNA sequence, we do not have a prior way to know which of the two strands will be transcribed, and then translated. Moreover, certain regions of DNA in many organisms are transcribed in both directions. Therefore, for a DNA sequence for which we want to assess its potential as a coding sequence, we must evaluate the possibility that transcription and translation occur on either or both strands. Thus, if we combine the existence of three reading frames for each strand with the evaluation of both of them, we see that our algorithm must translate the six potential reading frames from the original DNA sequence to be evaluated, three directly from the sequence, and another three from the computed reverse complement.

Having obtained the six possible reading frames, we must now think about how to make our algorithm determine if any of them has the potential to encode for a protein. To proceed, we will introduce the last concept of this section, that of the Open Reading Frame (ORF). To understand an ORF, let's briefly revisit our analysis of the genetic code and remember that out of the 64 possible codons, 3 are "STOP" codons. This implies that if one were to generate random DNA sequences, one STOP codon would appear on average every 21 codons ($64/3 = 21.3$). In contrast, proteins (or coding regions of exons in eukaryotic organisms) typically have significantly longer lengths, usually greater than 100 amino acids. So, once the amino acid/stop sequences are obtained in the six possible reading frames, what our algorithm needs to do is count, for each frame, the number of codons that code for any of the twenty amino acids that appear between two STOP codons, and within the same reading frame. Those regions, where between two STOP codons, there are more than, for example, 50 or 70 codons (potential amino acids), are considered open reading frames (ORFs), and correspond to those DNA sequence regions that are putatively coding for proteins (or exons). In conclusion, our computational algorithm comprises a series of steps that use specific rules (genetic code, base complementarity, threshold for defining an ORF) and operations on the sequence, and is able to determine from an "input" (initial sequence) the "output" of the open reading frames found in it, and its corresponding protein sequence. This example, encompassing a simple algorithm that the reader should be able to follow step by step, includes the fundamental elements of a typical bioinformatics algorithm—the "input," contextual information, underlying logic, and "output." This framework will serve us in the following sections and chapters to explore the bioinformatics algorithms used in clinical genomics. Through this short exploration, we not only glimpse the crucial intersection of biology and computer science but also appreciate the nuanced challenges that bioinformaticians encounter as they translate vast datasets and convert them into precise genetic insights. This algorithmic approach, foundational to the broader field of bioinformatics, illustrates the sophisticated infrastructures developed to interpret life at its molecular core.

The Two Fundamental Problems of Bioinformatics Applied to Genomics

There are two fundamental issues associated with the bioinformatic analysis of data derived from massive sequencing technologies, namely the reads. First, when the genome of a species is sequenced for the first time, the challenge is to assemble the whole genome from the reads. This process is known as "de novo assembly." Second, when the species genome is already known (and we refer to it as the reference genome), the challenge lies in comparing the reads derived from an individual's genome against this known reference genome (a process known as mapping), and subsequently determine the differences between them, which technically are referred as the variants.

To understand the complexity of both problems and begin to address them, let's first use an analogy that helps clarify the relationship between the sequencing reads obtained from a sample and the actual underlying genome sequence of that sample. When sequencing an entire genome, our goal is to uncover how the complete genome sequence relates to the many shorter subsequences that we observe/obtain as "reads." Imagine that the genome—the full sequence of A, C, T, and G bases—is like the complete text of a large book. Sequencing it using high-throughput (massive) sequencing technologies would be somewhat like this process: (i) first, we separate all the pages of the book so that they are individual sheets. (ii) Secondly, we cut each page horizontally into two or three large strips—this represents the initial fragmentation of the genome. (iii) Third, we make millions of copies of each of these strips using a photocopier; this step corresponds to the clonal amplification. (iv) After that, step four, we take the photocopied strips and cut them once more, this time into very small snippets—each perhaps a sentence or two long, containing around 100 letters. This can be considered as the potential reads. All these snippets are then thoroughly mixed together in a huge (really huge) bag. Finally, Step 5, we randomly draw millions of these small fragments from the bag—these are our sequencing reads. The challenge that follows is to reconstruct the original book—the complete genome—from this enormous, shuffled collection of overlapping fragments.

To better appreciate the relationship between the "reads" and the "genome" within the context of our analogy, let's delve into some numerology. A human haploid genome has about 3 Giga Bases, that's a 3 followed by 9 zeros. If we use A4-sized pages, one page holds about 3000 letters, so our genome would equate to a book with one million A4-sized pages (quite lengthy, right?). If this million-page book is photocopied (amplified) 1 million times, we would end up with something like one to ten trillion fragments (each 3000 letters long). If we now cut into sentences (reads) of about 100 characters, we see that for a human genome, the total potential reads with standard amplification is about 30–100 trillion (or more). As stated in the previous chapter, in a sequencing experiment, we can obtain about 100 million reads, indicating that we continue to sequence about one in every thousand possible reads. To put it another way, the reads always represent a minor portion (obtained at random) of the total sequence contained in the sample. This explanation underscores the inherent challenges and complexities in genomic sequencing and highlights the craftsmanship required in bioinformatics to weave together these fragmented threads into a coherent genomic narrative. Managed effectively, bioinformatics not only solves these foundational puzzles but also enhances our grasp of biological systems, fostering breakthroughs in genetic research and medicine.

The de novo assembly of a genome, as already described in the chapter about the human genome, where Celera's strategy was briefly presented. The mapping algorithm will be described in the next chapter in the context of variant calling. But before moving forward, we have to describe and properly understand two additional fundamental concepts associated with the bioinformatic results of sequencing reads processing, read depth and coverage.

The choice between reconstructing a genome from scratch or identifying specific deviations is dictated by the two fundamental problems of bioinformatics, as illustrated in Fig. 4.1.

Concepts of Read Depth and Coverage

Remember that the sequence (the genome) of the sample is copied numerous (millions of) times. This is why the reads read more than once each specific "letter/base" (each position) of the genome, and this is where one of the central concepts in the bioinformatic analysis of sequencing data emerges: the concept of read depth. Read depth is the number of times each position of the sample's genome is read. In other words, it counts how many reads feature a given position of the genome. Depth is a property of each genome position, obtained as a result of the bioinformatic processing of the reads. Since the reads are obtained at random, each position will be observed in more

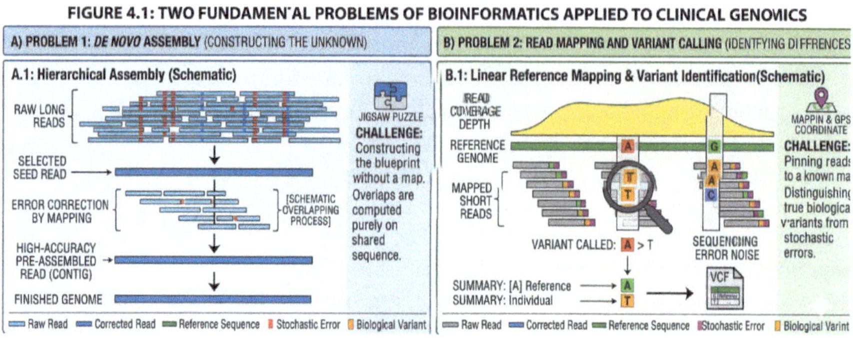

Fig. 4.1 The two fundamental computational problems in bioinformatics. The transition from raw sequencing data to genomic insight follows two distinct pathways: (**A**) De Novo Assembly. (**B**) Read Mapping and Variant Calling

or less reads depending on the sample's outcome. To obtain a single parameter that describes the experiment we can calculate, once we determine the read depth at each position, the average read depth value. The higher this value is, the more certainty there will be about the base at each position, and therefore the sequencing will be considered more precise. When designing a sequencing experiment, an estimate is usually made of the size of the target sequence. For instance, 3 Gb for a human genome. Before starting the sequencing, the user also decides how many reads to obtain and their length, so in the experimental design, it is decided how many total bases will be sequenced. For example, on an Illumina machine with 100 bp reads, if 300 million reads are to be obtained, the total of sequenced bases would be 30 Gb. The ratio between the total number of bases to be read (30 Gb), and the length of the target sequence (3 Gb) in this case is 10. This value is called the theoretical or expected read depth, and is the value referred to as "X" value when designing a sequencing experiment. In our example, the genome was sequenced at 10× (30 Gb/3 Gb). If the sequencing and the bioinformatic processing of the reads were correct, the theoretical (or expected) depth should be very similar to the average read depth obtained after processing the reads. Alternatively stated, if the average depth obtained matches the expected depth, we can be confident that the sequencing was successful. Finally, after processing the reads and calculating the read depth, the concept of coverage emerges. The coverage reflects what percentage of the desired sequence has indeed been read, at a certain quality threshold. Usually, the quality threshold is set as a minimum read depth. Thus, for example, let's say we set the threshold at ten minimum reads. The coverage will be the percentage of bases of the target sequence that have been read 10 or more times. In a typical example, a human genome can be sequenced at a theoretical depth of 30×, usually resulting in a threshold of ten reads, in a coverage exceeding 90%. The foundational concepts of depth and coverage provide a practical framework for understanding the precision and efficiency of genomic sequencing projects and the results of the subsequent bioinformatics processing. These metrics not only influence the design and expectations of sequencing projects but also ensure the reliability and replicability of genomic research, thereby serving as critical indicators of the quality and completeness of the gathered data.

In conclusion, bioinformatics is not merely a supplementary field, but a cornerstone of modern biological research and particularly genomics, including as expected clinical genomics. The sophisticated interplay of computational tools and biological data continues to illuminate pathways previously shrouded in mystery, transforming abstract genetic sequences into knowledge. As these technologies continue to evolve, they not only expand our understanding of the biological landscape but also enhance our ability to predict, diagnose, and treat diseases with remarkable precision. The ongoing development and integration of bioinformatics into the fabric of genomics heralds the advent of an era where computational prowess and biological knowledge converge for the betterment of human health. To begin this journey in the field of clinical genomics, let's move to the next chapter.

References and Suggested Further Readings

Sims D, Sudbery I, Ilott NE, Heger A, Ponting CP. Sequencing depth and coverage: key considerations in genomic analyses. Nat Rev Genet. 2014;15(2):121–32. https://doi.org/10.1038/nrg3642.

Lesk AM. Introduction to bioinformatics. 5th ed. Oxford University Press; 2019.

Chapter 5
From Genome to Variants and Their Meaning

Abstract This chapter describes the computational workflow that transforms sequencing data into clinically meaningful genomic insights. It begins with read mapping, the process of aligning millions of sequencing reads to the Human Reference Genome, highlighting the role of efficient algorithms such as the Burrows-Wheeler Transform in solving this large-scale problem. Once reads are mapped, variant calling identifies genomic differences between the sample and the reference, distinguishing true variants from sequencing errors using metrics such as read depth, allele balance, and quality parameters. The resulting variants are stored in standardized formats such as VCF files. The chapter then focuses on variant annotation, a multi-layered process that assigns biological meaning to each variant. Structural annotation determines genomic context and molecular consequence, while functional annotation integrates data from key resources such as gnomAD, UniProt, and OMIM to provide population frequency, protein impact, and clinical associations. Finally, the chapter introduces variant classification frameworks, emphasizing the identification of pathogenic variants within the vast background of benign variation. This process, often compared to finding a needle in a haystack, represents the core challenge of Clinical Genomics and establishes the foundation for clinical interpretation and diagnosis.

Keywords Read mapping · Variant calling · Variant annotation · VCF · Pathogenic variants

Mapping the Reads

When working with MSTs on patient samples, having sequenced their genome (whether whole genome, exome, or panel), it is crucial to remember that what is obtained are a multitude of reads (millions of them). Depending on the experimental design, these reads will have covered the target sequence several times (depth). Starting with these reads

M. A. Marti, *Clinical Genomics*,
https://doi.org/10.1007/978-3-032-34667-4_5

and the Human Reference Genome (HRG), the first step we must take is to determine where (i.e., at what position or coordinate) in the HRG each read corresponds. This process of situating the reads on the HRG is known as Mapping. At an initial stage, and to conceptually understand the mapping process, let's assume that the reads do not contain variants, meaning they have exactly the same sequence as the reference genome.

For example, suppose you have the following read: ACTGACTAGCATCTG. Mapping it means determining that this sequence in the HRG is, for example, on chromosome 12 starting at position 34,457,872, aligning it in this way:

Position on Chromosome 12:34,457,872 HRG: ACTATATCTGCTG TGCATCGTACGTACGTA ACTGACTAGCATCTG CAACGTGCGTAC Read: ACTGACTAGCATCTG

This problem is extremely complex for large genomes, like the human genome, especially with short reads. Why is that? In a typical case, suppose a Human Exome was sequenced with an expected depth of 100×, let's say the exome totals about 100 Mb (approximately 3% of the genome), which would yield a target sequence of 100 × 100 Mb = 10 Gb. If sequenced with a typical second-generation device like an Illumina that allows for 100 bp reads, this gives a total of 100 million reads that must be mapped. A read could "start" anywhere across the 3 Gb of the genome, meaning that if one examines them one by one, one would need to check 3G positions for each read, totaling 3G × 100M = 3 × 10^17, or a 3 with 17 zeros. If a computer takes just 3 milliseconds to check a position and see whether the read matches or not, it would take about 1 × 10^14 s to map all the reads, which is equivalent to roughly three million years. Clearly that is not useful. Fortunately, there is an algorithm that performs the mapping of the reads much more efficiently (in computational terms). This algorithm, called the "Burrows-Wheeler Transform (BWT)," is currently used by everybody for mapping reads to a reference genome. While we won't delve into its complexities, it's useful to briefly explain how it works. The BWT begins by generating an index of the human genome (similar to the index of a book). The index is made up of all possible subsequences of the HRG starting from each of its bases to the end. That is, if the genome is 3 GB, the first word is the entire HRG, the second is the subsequence starting from the second base going until the end; the third begins at the third base, and so forth up to the total possible words. This index is then alphabetically sorted. Each word is also associated with a number indicating the position where it begins in the genome. Thus, the first word corresponds to position (1), the second to position (2), and so forth. Here is an illustrative example:

Example of a mini HRG: ACTTGTAACTCT

Words in the index: ACTTGTAACTCT (1) CTTGTAACTCT (2) TTGTAACTCT (3) TGTAACTCT (4) .... CT (11) T (12)

Where the number in parentheses corresponds to the position in the HRG of the first letter of the word. Now after sorting them alphabetically, we obtain the index.

Alphabetical Word Index: AACTCT (7) ACTCT (8) ACTTG (1) .... T (12) TAACTCT (6) TT (3)

It is important to note that a read (if there are no errors or variants) is actually a sub-sequence of the genome, and therefore there will be a word in the index whose beginning matches exactly the read. In the illustrative example, if the read was "ACTT," you would see that it corresponds to the first word ACTTG (1). Conversely, the read "TAA" corresponds to the sixth word TAACTCT (6). With this procedure, it is easy to observe that if you want to know "where" a read is located, you should traverse the index until the read matches the beginning of the word. Once the word that begins with the read is identified, the number associated with the word, which corresponds to the position of that word in the genome, will also correspond to the position of the corresponding read in the genome, thus effectively mapping it. In our example, it's easy to see that the read "ACTT" should be mapped to position 1, while "TAA" maps to position 6, as shown below:

Position 1 6 Mini HRG ACTTGTAACTCT Read1 ACTT Read2 TAA

From this example, it becomes clear how the algorithm for mapping reads, only needs to traverse the index for each read until it finds the word in the index whose start matches the read. This method is quick and efficient for mapping reads, and as we will see next, it also allows for mapping reads with variants. It is important to note, that reads are not searched completely in the index, but incrementally from each of the ends. For example, suppose a read of 100 bp, finding the word that contains it in the index typically does not require matching all 100 bp, as with the first 10–20 the word in the index that continues it will have been found. This can easily be seen again in our example. Suppose we obtained the following read "ACTTGTAA". If we begin with the first letter, "A", we will see that in the index there will be several words that begin with "A." Indeed, there are three, the (1), the (7), and the (8) (they are the first three in the ordered index). If we now continue in the read and take two letters, that is "AC", and look in the index, we will see that there are two (1) and (8), the same happens if we take three letters "ACT". Now when we take four letters "ACTT," we see that in the index only the first word contains ACTT at its start, and therefore the read should be mapped to the position (1). To see if this was understood, you can try mapping the following reading "CTCT". Now suppose that we have a read similar to the previous one but with a variant in the sixth base of the read (we change a T>G). so the read with the variant is ACTTGGTA. If we map this read, as in the previous case, we can observe that with only the first four letters "ACTT," it already shows us that the read maps to position 1, remaining then the read with the variant mapped in this way:

Position 1 6 Mini HRG ACTTG**T**AACTCT Read wVar ACTTG**G**TA

The resulting mapping clearly shows the variant in position 6, as T>G. Mapping programs use one of the ends of the reads to map them, which allows the variants contained beyond the mapped end to be mapped/aligned against the HRG and later identified. It is also important to note that if a read has too many errors, it is possible that we will not find a word that contains it in the alphabetical index of the genome. In this case, the read **cannot** be mapped and will not be used in subsequent steps. Having seen how the readings are mapped onto the reference genome, we will now see how the variants are called.

Variant Calling

Once the reads are mapped onto the HRG, we must determine the variants present in the sample. This process is known as variant calling. Variant calling involves an algorithm with two components. The first, relatively simple step, involves traversing each of the mapped reads and comparing them against the reference genome, marking the observed differences. These differences are potential variants, and the second step of the algorithm consists precisely in determining which of the observed differences correspond to a real variant (present in the sample), and which are sequencing errors (i.e., the sample has the same sequence as the HRG). To differentiate between real variants and errors (also called false positives), the depth obtained at that position is primarily used.

Remember that in each of our cells the genome is duplicated (we are diploid organisms), and therefore, for each position in the genome (locus), we will have two alleles: one inherited from the mother, and the other from the father. Assuming that the sequencing process was correct, it is logical to assume that at a given position was read 100 times (i.e., its depth is 100), with ~50 reads (approximately) coming from the maternal allele, and the rest from the paternal allele. Now, if our mother at that position does not carry a variant, the 50 reads from the maternal allele would be identical to the reference genome, so the first step of the algorithm would not mark anything. Now, suppose that the father possesses in his two alleles (i.e., he is homozygous) a variant at that position (for example C>T), as a result, all 50 paternal reads will reflect this, and in all these reads, the algorithm will mark the observed difference. What we will then observe after the mapping and marking of differences in the position is that 50 reads (the maternal ones) show a C, just like the HRG. The other 50 reads, the algorithm will mark each with a T. When further processing this position, the "variant caller" sees that 50 out of 100 reads all have a T marked as a difference, it will indeed say that this is a real variant. In fact, being in "only" one allele, the algorithm will call the C>T variant as heterozygous. The same logic allows us to observe that if the depth at a given position was sufficient (for example more than 30) and in approximately half of the reads we consistently observe the same difference, the corresponding variant will be called as a heterozygous variant. It's easy to see that if "both" alleles in the sample present the variant, then it will be observed and marked in all (or almost all the reads) consistently, and consequently, the program will call the variant and in this case as a homozygous variant (i.e., it is present in both alleles).

So, what is a false positive in this context, or in other words a difference noticed in the read relative to the HRG, but where a variant not called? -Suppose that at a given position of the sample it does not possess a variant, i.e., in both alleles the base corresponds to the expected base in the HRG, suppose in our example an "A". In this case, all reads should have an A at that position. Now, if in any of the 100 reads there was a sequencing error, and for example a "C" was read, it will be marked as a difference. But it will be a "C" among 99 "A"s, so clearly, it will not be called as a variant. Even if the error rate of the equipment were an extremely high 10%, the error should be random, so the nine to ten erroneous bases should be distributed among C, T, and G, and none exceeding three to five reads. In other words, we expect sequencing errors to be isolated differences, while real variants are recurrent differences, thus observed in a significant number of reads. In the Figure below we show an image illustrating the mentioned concepts.

Figure Showing Mapped Reads, Marked Differences and Variants

HRG:ATGTACTCTACGTTACGCCATCATCAGCATCACACGTACATCGA
TGCTATTCT reads: ATGTGCTCTACGTTACGCCACCATCAGCTTCACA
TCTACGTTACGCCACCATCAGCATCACACGTATAT
CGCCACCATCAGCTTCACACGTACATCGATGGTA
GCCACCATCAGCATCACACGTACATGGATGCTACT
TTCGTTACGCCACCATCAGCTTCACACGTACA
ATGTACTCTACGTTACGCCACCATCAGCATCACA
GTACTCTACGTTACGCCACCATCAGCTTCACT
ATGTACTCTACGATACGCCACCATCAGCAT
TCTACGTTACGCCACCATCAGCATCACACGTACAT
CGGCATCATCCGCATCACTCGTACATCGATGCTA Variants C T

The algorithms that call the variants analyze not only the depth of the difference but also other parameters associated with the sequencing process. These parameters include the base quality (i.e., how confident the equipment is that the base it read is indeed that base), the position of the base within the read, and what is called "strand bias." When sequencing occurs normally, since fragments are randomly selected, it is expected that half will correspond to sequences of one strand, and the other half to those of the complementary strand. After mapping, in addition to depth at each position, it is possible to measure the ratio between the amount from each of the strands of the GHR to which the read corresponds. This value should be close to 1, and the further it deviates from this value, the more strand bias the position will exhibit. Positions with a strand bias far from 1 indicate a potential problem in the sequenced region and possible errors in variant calling at that location.

To calibrate all these parameters and obtain the quality thresholds for the variants, the algorithms often use a procedure called VQSR (Variant Quality Score Recalibration). In this procedure, all parameters for all observed differences are first determined. Then, using a Database of common variants in the population, it is determined which of the differences correspond to variants already observed in the population and are therefore considered real variants present in the sample. For this set of real variants present in the sample, the mean and standard deviation are determined for each type of variant (homozygous or heterozygous) and for each parameter. With these values, novel real variants are defined as those whose parameter values fall within 2 (or 3) standard deviations from the mean. On the other hand, differences whose values, for instance of strand bias or depth, are greatly deviated from what is expected will be considered sequencing errors. The VQSR step typically discards between 10%–20% of the observed differences as sequencing errors, keeping only those quality variants.

The result of the variant calling is then a file, called a VCF (Variant Call Format), which contains the list of variants that characterize the sample. The VCF file will subsequently be used in the process of assigning potential biological functions to the variants, in what is called the annotation process. Following variant calling, the file contains a row for each variant with basic information about it in relation to the HRG and its quality. This information includes:

The position in the reference human genome in Chrom:Pos format, which indicates the Chromosome and position (i.e., 1:346565222). The nature of the variant, such as G>T, indicates that it is a variant involving a single nucleotide substitution from G to T. The depth of each base, and the total depth, for example, 46:52:98, which indicates that G was read 46 times, T 52 times, totaling 98, implying that the variant is in heterozygosis. At this point, the VCF also contains other parameters associated with the quality of the variant related to the variant calling process.

Variant Annotation

Having obtained (called) the variants, the next step is to determine their biological significance. To do this, the first step is to add (or associate) biological information to each variant that must later be interpreted by the analyst. The process of associating biological information to each variant is known as variant annotation, and in the jargon, it is said that the variants are annotated. The annotation process is divided into multiple steps or layers, which annotate and thus add different types of information. The first step, which is the most standardized, is known as structural annotation, and it is based on using the HRG as a map.

Remember that the HRG is like a map, containing frontiers, rivers, and cities, which in our case are the genes, their exons, introns, regulatory areas, etc. Thus, the first annotation each variant receives is related to its position on the map. It is analogous to saying, given a coordinate in space, that it is located in a certain country, province, city, etc. Therefore, the structural annotation of the variant will tell us in which gene it is located (or if it is in an intergenic region), and within the gene if it is in an exon, an intron, a promoter zone, etc. Second, given the nature of the variant (for example C>T), we can determine its molecular effect. The molecular effect can first be classified according to whether the variant is located in a coding region (an exon) or not. If it is an intronic variant, it will be determined if it could affect (or not) splicing, and/or if it is located in a regulatory area. If the variant is in a coding region, it can be a synonymous variant, which is when due to the degeneration of the genetic code, the base change does not produce a change in the resulting amino acid. Alternatively, the variant can be a missense variant, which occurs when the resulting base change does result in an amino acid change in the protein. In this case, the variant will be annotated with the corresponding change in the protein, and referred to its position within it. For example, p.R343T means that in the resulting protein, a change from Arginine (R) to Threonine (T) will occur at residue 343. Finally, the variant can be a nonsense type of variant, when the change produces a STOP codon, and/or a change in the reading frame (also called a frameshift variant).

Structural annotation allows us a first approximation of the relevance analysis of the variants. It is clear that if we rank them in terms of their potential biological impact or effect, the nonsense variants and those that affect splicing will be variants with high potential impact, possibly having a significant biological effect. At the other end, synonymous variants, and those intronic and/or in intergenic regions probably have a low or null impact, since they will most likely not affect the biology of the individual significantly. Finally, missense variants will surely require a more in-depth analysis involving estimating or determining the potential effect of the residue change in the resulting protein.

The second step of the annotation, usually called the functional annotation, involves associating each variant with general biological information about it, the gene that contains it, and/or similar variants. This process is much less systematized than structural annotation and is where the difference between a good and not as good analysis protocol often lies. Among the most important functional annotations of a variant, we have those related to population genomics, i.e., its allelic frequency in different populations; those related to its molecular effect (beyond the structural annotation), such as specific knowledge about the role of the affected residue in the protein, knowledge of the effect of other variants in the same gene and related functional data. And finally, and extremely important in Clinical Genomics, information related to known effects of variants in the same gene concerning clinical aspects, such as symptoms observed in other patients. Various biological databases that contain the association between the gene/variant and biological effect are typically used to perform functional annotation.

Currently, for example, everything related to the population frequency of variants is usually derived from the gnomAD project (Genome Aggregation Database), a comprehensive and immensely valuable resource dedicated to the aggregation and analysis of genomic data. The project launched in 2014 as ExAC is developed by the Broad Institute, aims to compile data from exome and complete genome sequences from various human populations, currently reaching nearly one million samples. GnomAD provides an accessible and public platform that allows researchers to study genetic variation across different ethnic groups and better understand the relationship between genetic variants and diseases. The database is continuously updated, incorporating new sequences and refining data quality, facilitating the discovery of rare and common genetic variants and their impact on human health.

For the determination of the molecular effect in the protein, the best resource is UniProt (Universal Protein Resource), a comprehensive and open-access database that provides detailed information about the function, structure, subcellular location, and post-translational modifications of proteins, as well as links to other databases and relevant scientific literature. The information can be analyzed in a residue-specific context, particularly knowledge about the known variants therein and their molecular effect, including information from functional studies of different mutants. This database facilitates research in molecular biology, genetics, and medicine, helping to better understand proteins and their role in health and disease.

Finally, concerning the clinical aspects associated with the observed symptomatology (phenotype), the database par excellence is OMIM (Online Mendelian Inheritance in Man), a comprehensive and manually curated database containing detailed information about human genes and their genetic diseases. Created by Victor A. McKusick and maintained by the U.S. National Center for Biotechnology Information, OMIM provides clinical descriptions, genetic and molecular mapping data, and links to relevant scientific literature. Additionally, it includes information about specific genetic variants and their relationship with particular phenotypes, facilitating the identification of genes associated with hereditary diseases. This database is fundamental for genetic diagnosis, researching disease pathogenesis, and developing new therapies.

Besides the mentioned databases, there are numerous other databases that allow annotating the variants with biologically relevant information, specific to certain pathologies such as cancer, and/or certain aspects of clinical genomics like pharmacogenomics, and risk assessment, among others.

An Initial Approach to Variant Classification

Once the variants are annotated, the goal now is to determine which ones are relevant to health, that is, those that have clinical relevance. Within the context of clinical genomics, variants are classified based on their annotations and according to the criteria of the American College of Medical Genetics (ACMG) into five categories: Pathogenic, Likely Pathogenic, Variant of Uncertain Significance, Likely Benign, and Benign. Beyond the details of how they are classified and the strict meaning of each category, which we will see later, at this point, it is important to highlight that what we are looking to determine are the (Likely) Pathogenic variants, since the criterion of pathogenicity implies that possessing the variant contributes to the development of a pathology (hence the term), and thus affects our health.

The question now should ring in the reader's mind is: What variants will conceptually be pathogenic, and what characteristics do they possess? A pathogenic variant is one that, due to its molecular effect on the gene/protein, results in a pathological phenotype. It is clear, then, that those variants expected to have a high molecular impact (for example, the nonsense variants) are more likely to be pathogenic. Therefore, the ultimate goal of the annotation is to allow the analyst to analyze and evaluate the variant information, its annotation, to firstly determine its pathogenicity and secondly, its clinical relevance. The problem is that out of the millions of variants each human genome possesses, only one (or a few) are clinically relevant. This is why the process of finding a clinically relevant variant in a genome is often likened to finding a needle in a haystack. Beyond this difficulty, it is important to emphasize that better annotation, which is the bioinformatics' objective, will allow for more precise classification, and facilitate the search for clinically relevant variants.

Towards Clinical Genomics

In this final section, let's revisit the concepts addressed in the last three chapters to see how they integrate to create clinical genomics. Firstly, we have the Massive Sequencing Technologies that allow for efficient DNA sequencing of a sample; in the particular case of clinical genomics, this will comprise the patient's genome (or a fraction of it, such as the exome or a gene panel). The result of the sequencing process is a set of short sequences that we call reads. Taking the Reference Human Genome, which is a map

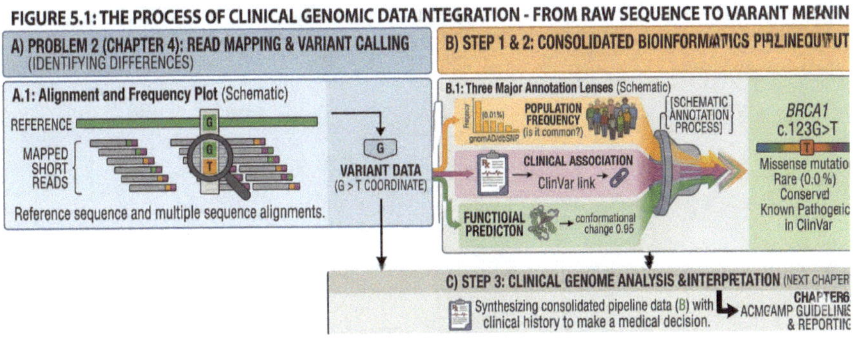

Fig. 5.1 From raw sequence to variant meaning. (**A**) Mapping and Variant Calling. (**B**) Variant Annotation. (**C**) Clinical Interpretation

of our species' genome, as the starting point, the process will firstly locate the reads on the HRG, that is, perform the read mapping, and then determine where the patient's genome differs from the HRG. These differences are the variants. At this step, we can state that the patient's genome can be represented as a list of variants, contained in a file called the VCF. Once the VCF is obtained, biological information will be added to the variants, first using the reference genome as a map (structural annotation), and then associating each of the variants with different biological processes (functional annotation). Finally, based on the annotations, the analyst will classify the variants and seek those that have the capacity to affect the patient's health, that is those that are (likely) pathogenic, and are important for the clinical picture, i.e., have clinical relevance.

Summaryzing, the transition from a raw VCF list of coordinates to an informative medical dataset requires the complex bioinformatics annotation pipeline shown in Fig. 5.1, which converts simple single-nucleotide disagreements into contextualized clinical findings.

In the next chapter, we will first address the historical cases of clinical genomics in the context of diagnosing rare diseases and then systematize the analysis of cases and classification of variants.

References and Suggested Further Readings

Heng L, Richard D. Fast and accurate short read alignment with Burrows–Wheeler transform open access. Bioinformatics. 2009;25(14):1754–60. https://doi.org/10.1093/bioinformatics/btp324.

Koboldt, DC. Best practices for variant calling in clinical sequencing. Genome Med 2020;12:9. https://doi.org/10.1186/s13073-020-00791-w.

ilalis E, Tavakoli N, Razaghi-Moghadam Z., et al Genome-wide functional annotation of variants: a systematic review of state-of-the-art tools, techniques and resources. Front Pharmacol 2025;16:1474026. https://doi.org/10.3389/fphar.2025.1474026.

Austin-Tse CA, et al Best practices for the interpretation and reporting of clinical diagnostic WGS. npj Genomic Med 2022;7(1):27. https://doi.org/10.1038/s41525-022-00295-z.

Chapter 6
Application of Clinical Genomics to the Diagnosis of Mendelian Diseases

Abstract This chapter explores the application of Clinical Genomics to the diagnosis of Mendelian diseases, highlighting how massive sequencing technologies have transformed medical genetics into a data-driven, genome-centered discipline. It begins by reviewing Mendelian inheritance patterns-autosomal dominant, autosomal recessive, X-linked, and de novo-and explains how each model constrains the expected characteristics of pathogenic variants. The chapter then examines the molecular mechanisms underlying disease, including loss of function, dominant negative effects, and gain of function, emphasizing their implications for variant interpretation. It further discusses the landscape of disease-associated genes, noting that only a subset of human genes contribute to Mendelian disorders due to biological tolerance or essentiality. Through landmark case studies, including early exome-based gene discovery and the first successful genomic diagnoses, the chapter illustrates how variant filtering, annotation, and inheritance modeling enable the identification of causative mutations. These examples demonstrate not only the feasibility of genomic diagnosis but also its direct clinical impact, including targeted therapeutic interventions. Finally, the chapter reflects on the broader significance of Mendelian diseases as natural models for understanding gene function, positioning Clinical Genomics as a cornerstone of precision medicine and translational research.

Keywords Mendelian diseases · Inheritance patterns · Pathogenic variants · Clinical diagnosis · Precision medicine

One area that emerged in the last decade as one of the first beneficiaries of MST (Massive Sequencing Technologies) was medical genetics, and as a result, it is now inconceivable to consider medical genetics (pardon the redundancy) without clinical genomics. Medical genetics is the branch of medicine that traditionally focuses on the study of genes and their impact on human health, encompassing the research and diagnosis of genetic and hereditary diseases, the main representatives of which are the Mendelian diseases. In

M. A. Marti, *Clinical Genomics*,
https://doi.org/10.1007/978-3-032-34667-4_6

this context, the quantum leap from Sanger-type sequencing to MST resulted, in practice, in an almost complete transformation of medical genetics into medical genomics. Or directly into clinical genomics, although, as we have seen and will develop further, it also encompasses other areas of medicine. The goal of medical genetics is to determine which (or which ones) of the variants present in a patient are responsible for the observed pathology, or in strict terms, for the pathological phenotype. To do this, all the steps developed in the previous chapters must be followed: sequencing, mapping against the reference genome, calling and annotating the variants, so that the analyst can then prioritize them, and finally determine which one has the ability to explain the observed pathology. But what characteristics must these variants have in order to explain the pathology? That is, are pathogenic? How do they relate to being responsible for Mendelian diseases? What genes contain these variants, and how were they discovered? And finally, what does a molecular diagnosis of these pathologies using MST entail?

Mendelian Diseases

Mendelian disorders are those that follow a simple Mendelian inheritance pattern, meaning their appearance in the family tree (pedigree) conforms to what Mendel observed with his peas and is embodied in the famous Mendel's laws. Genomically, these diseases are usually monogenic, which means they are caused by the malfunction of a single gene. In terms of the variant's effect on the phenotype, it is said that the effect size (odds ratio) is infinite. They present a completely deterministic view of the genotype-phenotype relationship, since possessing the "pathogenic" genotype, inevitably leads to the development of the pathology. These diseases can be classified into different categories according to the mode of inheritance, including traditional autosomal recessive, autosomal dominant, and sex-linked diseases. In clinical genomics, as we will see below, we will expand, or specify this spectrum a bit more, including those that follow a "de novo" model. Let's take a moment to review each of these categories from a classic genetics perspective. Autosomal recessive diseases, such as Tay-Sachs disease or cystic fibrosis, require an individual to inherit two defective copies of the gene, one from each parent, for the disease to manifest. In contrast, autosomal dominant diseases, such as Huntington's disease or familial hypercholesterolemia, manifest when a single defective gene copy is sufficient to cause the disease. Sex-linked Mendelian diseases, such as hemophilia A or color blindness, are caused by genes located on the sex chromosomes, generally the X chromosome. Since men have only one X chromosome, the presence of a defective gene on that chromosome is sufficient for them to develop the disease. In women, who have two X chromosomes, the disease only manifests if both X chromosomes carry the defective gene. Finally, "de novo" diseases are those that occur when neither parent nor any ancestors has the disease, and it appears spontaneously in one of the children. In contrast, diseases with non-Mendelian genetic associations, which present complex inheritance patterns and often have a multigenic origin, will be analyzed later. Let's now analyze each type of disease, focusing on what characteristics the variants must have in each case.

Dominant Autosomal Diseases

In a dominant autosomal disease, the pathology will manifest when the individual has (at least) one copy of the pathogenic allele. In other words, the pathogenic variant that explains the case is present in the sick individual in heterozygosity. In such cases, a sick parent, for example, will statistically speaking have half of their children healthy and the other half sick; which means that an ill individual has a fifty percent (50%) chance of having a child with the same pathological phenotype. In the pedigree, the autosomal dominant pattern is evident because there are usually sick individuals in all generations. When a case shows that a sick parent has a healthy child, all descendants of this child will be healthy. This is easily explained from a genomic perspective, as what has happened is that the healthy child has inherited the healthy allele from the sick parent, meaning the variant is absent, and therefore will not reappear in that lineage anymore. From a population frequency perspective, the variants responsible for this type of pathologies tend to have a very low frequency, usually less than 0.1%.

From a molecular perspective in relation to the etiology of the pathology, there are different mechanisms. The simplest mechanism is due to what is called haploinsufficiency. This situation occurs when a single copy of the healthy gene is not enough to produce the necessary amount of gene product to maintain normal functioning. In these cases, having only one healthy allele is not sufficient for the individual to develop normally, and the pathology occurs. In these cases, heterozygous individuals present the classic pathology, and in the case of homozygous individuals, a different and usually more severe phenotype may be presented. In these cases, the causal variants of the pathology are usually loss-of-function type of variants, meaning they generate a non-functional allele. An example of a disease caused by haploinsufficiency is familial hypercholesterolemia, where the reduction in the number of LDL receptors (low-density lipoproteins) causes elevated blood cholesterol levels.

In other cases, the underlying mechanism is called "dominant negative." In this case, the variant produces a protein that is not only usually non-functional, but also interferes with the function of the normal protein produced by the non-mutated allele. This is why when the allele is present in heterozygosity, pathological development occurs. An example of this effect can be seen in osteogenesis imperfecta, where abnormal collagen proteins negatively interact with normal ones, compromising bone tissue integrity.

A third mechanism, which sometimes mimics the previous one, is the gain of function. This mechanism implies that the variant produces a gene/protein with a new function not observed in the normal protein and that results in pathological development. Often, the gene/protein becomes hyperactive due to the presence of the variant. An example of this case is Huntington's disease, where the pathogenic variants of the huntingtin protein acquire a toxic function that damages nerve cells.

This variety of molecular causes makes it very difficult to predict the effect of novel variants in this type of pathology, especially if the mechanism of pathological development is unclear.

Diseases Linked to the X Chromosome

X-linked diseases are caused by pathogenic variants in genes located on the X chromosome. Due to differences in the number of X chromosomes between men and women, these diseases have unique inheritance patterns and affect both sexes differently. The most common case includes those that are X-linked and recessive. In this case, men (who have a single X chromosome) will manifest the disease if they inherit an allele with the pathogenic variant. Women (who have two X chromosomes) will need to inherit both pathogenic alleles (one from each parent) to manifest the disease, which is less common. However, a woman with a single pathogenic allele will be a carrier and may pass the mutated allele to her children. Consequently, in the pedigree, it is observed that sick parents always have healthy children. The male children, not inheriting the X chromosome from the father, will not only be healthy but also do not possess the pathogenic variant, which means the disease will not manifest in future generations. In contrast, female daughters will all be carriers of the variant (on one of their two X chromosomes), and therefore their male children will have a 50% chance of inheriting the variant and thus developing the pathology. From an allelic frequency perspective, the variant will have a low or very low frequency, as we will see for recessive diseases in general. In these cases, the molecular mechanism usually corresponds to a loss of function.

There are also dominant X-linked diseases, although they are less frequent. In these cases, the pattern combines the complexities of an X-linked inheritance with a dominant-type disease as previously described. Nevertheless, in both cases, and as the name suggests, the key to an X-linked disease is that the pedigree shows a clear difference between the fraction and phenotype of affected men and affected women. On the other hand, always in these cases, the variant must be located on the X chromosome.

An important additional aspect to consider in these cases is observed when analyzing the depth of the X chromosome variants in affected men. As these individuals have only one X chromosome, and the Y chromosome only contains a few genes analogous to those on the X, males are haploid for the X chromosome. Consequently, a variant present on the X chromosome will be observed in all reads. In other words, in males, the variants of the X chromosome will always be called as homozygous (as they will appear in all the reads) from a variant-calling perspective.

Recessive Autosomal Diseases

These diseases manifest only when the affected individual possesses both pathogenic alleles. In these cases, if the parents are healthy, they must have each contributed one pathogenic allele, which implies that both are heterozygous for the pathogenic allele. This condition, having one healthy allele and one pathogenic allele, is referred to as being a healthy carrier, since the individual "carries" the allele capable of producing the disease without showing a pathological phenotype. Therefore, on the pedigree, it is observed that from two healthy parents, a sick child appears (with a frequency of 25%).

Often, pathogenic alleles pass through the family without ever encountering two carrier parents, so the pathology may appear spontaneously in families that were previously thought to be healthy. In other cases, the continuous presence of pathogenic alleles in the family, especially in communities where the pathogenic allele has a moderate frequency, results in the recurrence of the pathology in one or both parents' families.

From the perspective of the variant(s), we must distinguish two types of cases. In the first case, let's assume that both parents are carriers of the same pathogenic variant since, for example, it is a variant of moderate frequency that circulates in the population, or more commonly both parents have some level of consanguinity (are cousins, second cousins, or have a common ancestor). In that case, the affected child will be homozygous for the pathogenic variant, while each parent will be heterozygous for the pathogenic variant.

In the second case, let's assume that both parents possess pathogenic alleles, but the responsible variants are different. In these cases, the affected individual will be heterozygous for each of the two pathogenic variants. This condition is called compound heterozygosity and requires the affected individual to exhibit two pathogenic variants, both in heterozygosity and in the same gene. Since each variant is inherited from a parent, the picture is completed with each of them being a carrier of one of the two variants in heterozygosity and is necessarily homozygous for the reference allele at the position of the other.

From a molecular perspective, recessive autosomal diseases are often due to variants that cause a loss of function of the gene/protein containing them. To analyze the expected frequency, we must highlight an additional feature of these alleles: in some cases, healthy carriers may have increased resistance to certain infectious diseases. This effect causes an increase in the frequency of the pathogenic allele in regions where the infectious agent is endemic, resulting in a higher prevalence of the pathology. A classic example of this phenomenon is sickle-cell anemia. This disease causes red blood cells to adopt an abnormal shape, which can obstruct blood vessels and cause anemia due to the rupture of red blood cells. Sickle cell anemia is caused by a mutation in the hemoglobin gene, where adenine (A) is substituted by thymine (T), resulting in the substitution of the amino acid glutamic acid for valine. Hemoglobin is the protein responsible for transporting oxygen from the lungs to the tissues and is the main component of red blood cells. Healthy carriers have a mix of normal and mutated hemoglobin, but their red blood cells maintain a normal shape. Therefore, they do not present symptoms of anemia. However, when these heterozygous individuals are infected by the parasite that causes malaria, their red blood cells break more easily, preventing the parasite's reproduction and conferring resistance to malaria. In other words, heterozygous individuals have normal red blood cells, do not present anemia, but have increased resistance to malaria. For this reason, in areas where malaria is endemic, such as Central Africa, the Middle East, and India, the frequency of sickle cell anemia is significantly higher than in other regions, as carriers of the pathogenic allele have a selective advantage, and therefore, the allele is maintained in the population. The fact that there are healthy carriers, who may even have certain advantages in specific environmental conditions, results in the variants of recessive autosomal diseases having a moderate frequency (even close to 1%), and often higher than what is observed for pathogenic variants of a dominant autosomal disease.

De Novo Diseases

De novo diseases result from novel variants that are not present in the parents of an affected individual. These variants typically appear for the first time in the egg, sperm, or developing embryo. Due to their spontaneous nature, de novo diseases can arise in any family, with no prior history of the disease. They also tend to occur in genes essential for cellular functioning, which can lead to a wide variety of clinical manifestations, from intellectual and developmental disabilities to specific syndromes with unique physical and metabolic characteristics. A notable example of a de novo disease is Rett syndrome, a neurological disorder that primarily affects girls and is due to spontaneous mutations in the MECP2 gene. Since most cases of these pathologies are highly debilitating, affected individuals are usually infertile, either physiologically, or because even having functional sexual organs and gametes, their physical and/or intellectual disability does not allow them to reproduce. Therefore, these diseases do not recur within the family and appear spontaneously. From a molecular perspective, the underlying mechanisms are similar to those seen in dominant autosomal diseases, and from a Mendelian perspective, they often occur under this inheritance model, requiring the presence of only one pathogenic allele for the pathological development to occur. As the individuals are sterile and the variants arise de novo in each afflicted individual, there are, in theory, no healthy individuals with the variant, so the frequency of the variant in databases of healthy individuals should be strictly zero. The affected individual will then usually be heterozygous for the pathogenic variant, and both parents must be homozygous for the reference at the corresponding position. In other words, they must not carry the variant.

The Genes Responsible for Mendelian Diseases

A frequently asked question is, how many possible Mendelian diseases exist? Given that ultimately a Mendelian disease is caused by the failure of a single gene, and that currently even those that can be caused by mutations in several genes of the same metabolic pathway or enzymatic complex are usually subclassified according to which gene possesses the pathogenic variant. Examples of this case include oculocutaneous albinism (OCA), where, depending on which gene the pathogenic variant is found, it is called OCA1A, OCA1B, OCA2, etc. It is reasonable to state that there are as many Mendelian diseases as there are genes that cause them. But therefore, the question becomes, how many of the human genes are the cause of Mendelian disease? While it is difficult to give a precise answer, and in recent years we have seen a continuous growth in our knowledge of genes responsible for Mendelian pathologies, the total number seems to be nearing an equilibrium, at a value of around 8000 genes. But why, if the RHG contains about 20,000 (or a little more) genes, why only about 8 (or maybe 10) thousand of them are responsible for a Mendelian disease?

On the one hand, we must highlight that there are genes that are completely tolerant to a loss of function. In other words, even if one possesses two non-functional alleles in one of these genes, no Mendelian disease develops. Examples of these genes are polymorphic and redundant genes such as olfactory receptors. In humans, there are about 400 functional olfactory receptor genes, and more than 600 pseudogenes. The loss of function in many of these genes does not significantly affect our overall ability to detect odors due to their great redundancy, and because the sense of smell is not critical for survival, its loss does not result in a pathology. Another example is the gene that forms the AB0 blood group system. This gene has three alleles, the A produces an enzyme that adds N-acetylgalactosamine to the H antigen, forming the A antigen. The B allele produces an enzyme that adds galactose to the H antigen, forming the B antigen. The 0 (zero) allele is a non-functional version of the gene, whose protein product is unable to modify the H antigen, leaving it unchanged. Individuals with genotype 0 (with two non-functional alleles) do not present A or B antigens on their red blood cells, and although this can affect transfusion compatibilities, it has no significant adverse effects on the everyday health of the individual, meaning its loss does not produce a Mendelian disease.

On the other hand, there are those genes that are so essential for cell viability, such as those encoding for the ribosome, or the DNA polymerase, that their loss of function results in an unviable zygote and therefore no individual. Some may allow the formation of gametes but are lethal at some stage of embryonic development, so they do not result in live births. These genes are therefore completely intolerant to loss of function (they are also called essential genes) and thus, by not even giving rise to an individual, are not responsible for Mendelian pathologies. It is then those genes that due to the loss (or gain of function) result in a viable but sick individual that lead to the development of Mendelian diseases.

The fact that there are many genes, just under 10 thousand, capable of producing Mendelian diseases, means that although each of these pathologies is a Rare Disease, usually with a prevalence lower than 1 per 2000 live births (and usually much lower, with typical prevalences of 1 per 100,000 cases), they collectively impact close to 5%–10% of all live births and represent up to 20% of infant mortality, and more than half of the causes of all pediatric hospitalizations, so they clearly have a huge impact on our health as a whole.

Let us now proceed, in this context, to see from a historical perspective, first how MST enabled the systematic identification of genes responsible for Mendelian Diseases, and second, to diagnose in detail at the molecular level patients affected by them, even going so far as to propose and advance effective therapeutic alternatives, with a bit of luck.

The Proof of Concept

In classical genetics, discovering genes responsible for Mendelian diseases required extensive and complex study of numerous healthy and sick individuals across multiple generations in affected families. This approach involves tracking the inheritance of molecular markers (like those used to assemble the Human Genome) along detailed family trees, performing laborious linkage and molecular genetics analysis. However, the arrival of MST revolutionized the field by offering a more efficient shortcut for the discovery of genes associated with these diseases. The first notable report of the use of these technologies dates from 2009, when a group of researchers led by Dr. Jay Shendure, from the Department of Genome Sciences at the University of Washington, succeeded in identifying the gene responsible for Freeman-Sheldon syndrome. This advancement was significant because the researchers were able to determine the causative gene from the genomic information analysis of just four unrelated affected individuals. Freeman-Sheldon syndrome is an autosomal dominant disease characterized by characteristic facial and skeletal abnormalities. The underlying cause is a non-functional allele in the MYH3 gene. This gene was previously identified using classical genetics techniques.

The study was based on sequencing the exomes of four individuals from different families diagnosed with Freeman-Sheldon syndrome. To analyze the variants, eight control (or healthy) individuals who had donated their samples to the HapMap project, and whose variants had been identified by microarray technologies, were also sequenced. It is interesting to note that the initial version of the exome was sequenced with a targeted depth of 100×. However, only 49% of the reads could be mapped, resulting in an average depth of 50×, and a coverage of the "target" regions of 96.3%. The work also sought to validate the mapping and variant calling protocols from exome reads, so initially, the variants called for the eight control samples were compared with the variants that had already been identified in them in the HapMap project. The comparison showed 99.9% concordance for homozygous variants and 99.6% for heterozygous, demonstrating that the entire process from sample collection to variant calling was effective and precise.

Having accurately obtained the variants, the next step was to annotate them, and then based on the analysis, (re)discover the gene responsible for the Freeman-Sheldon Syndrome. For this purpose, the researchers conducted the following analysis at the gene level. First, they retained all those genes where the four affected individuals had at least one variant that could affect the functioning of the gene/protein. In this initial stage, they retained the nonsense and missense variants and those (if any) that would affect the canonical splice sites. This resulted in about 2500 possible genes, which is an improvement from the total number of human genes (about 20,000), but still not enough. To reduce the list further, they eliminated all variants that had been previously found in genomes of healthy individuals. This is justified because, being a dominant disease, any variant that generates a non-functional allele would result in an individual with the pathology. This is equivalent to filtering currently by a gene with zero population frequency, absence of homo or heterozygous individuals in population genomics databases. With this filter, the researchers were left with only one gene, which was indeed the aforementioned MYH3. It is important to note that this work did not discover a new gene responsible for a Mendelian disease, as it had already been found by classical genetics; however, it was the first conceptual demonstration that genomic analysis based on MST could be used for this purpose in diseases whose origin was unknown. With this in mind, the same research group set out to search for a new gene the following year.

The Test Case

The chosen disease for the search for the causative gene was Miller syndrome, a Mendelian disease suspected of a recessive inheritance model, and whose causing gene (or genes) were unknown. It is important to note that Miller syndrome, like the Freeman-Sheldon Syndrome, presents a clinical manifestation involving various malformations, which together allow for a differential clinical diagnosis with high certainty. Miller Syndrome, also known as Genee-Wiedemann syndrome, corresponds to MIM263750 in OMIM "POSTAXIAL ACROFACIAL DYSOSTOSIS; or simply POADS." At the time of the study, the literature had described only three families with multiple cases, typically two affected siblings, usually children of healthy parents with no relation to each other. To search for the gene responsible for POADS, as in the test case, researchers determined the genome of four affected individuals and looked for genes that presented in all cases previously unreported variants that produced a change in the protein.

The analysis was much more complex than the previous case. Initially, the researchers considered not only a recessive model but also a dominant inheritance model, which complicates and multiplies the number of genes to analyze. Secondly, it is important to note that when selecting genes that could account for the disease in a recessive model, one variant in homozygosity or two in heterozygosity (compound heterozygosity model). Those variants should also not be found in control individuals, and should have a significant effect on the gene or protein. This model did not determine a single candidate gene but a dozen. And this number is even greater when considering a dominant inheritance model. Therefore, researchers also used an *in-silico* predictor (polyphen2) to evaluate which of the missense variants had a deleterious effect.

The analysis finally allowed them to identify a candidate gene, called DHODH, which codes for a protein that participates in the synthesis of DNA bases. It is interesting to note that for this gene, the model was indeed recessive having found six different variants in the three studied families, always in a model of compound heterozygosity, and where five of them were predicted as deleterious. The only one not predicted as deleterious by Polyphen-2, was a variant that changed a glycine to an alanine, a change which is usually difficult to predict the effect due to the intrinsic characteristics of glycine, the only non-chiral amino acid and which is highly flexible in relation to the secondary structure of the protein.

The confirmation of the responsible gene was carried out by sequencing, in another four unrelated cases with the syndrome, only the DHODH gene (using the Sanger method), and verifying the presence in all cases of at least two heterozygous variants with a possible deleterious effect on the protein. In total, 11 new mutations were found, ten of which correspond to SNVs that "simply" change an amino acid, while the remaining one is a deletion that generates a shorter, truncated, and non-functional protein. A notable point in relation to the found variants is that the classic change of R for W (or vice versa) is observed. Arginine (R) and Tryptophan (W) are perhaps the two most different amino acids possible, the former being the most polar and charged, and the latter an aromatic amino acid, but whose codons differ only by one base, hence this variant frequently appears and is usually pathogenic. Another interesting point is that the other pathological variants observed occur due to losses of Cysteines, or glycines, residues whose change tends to alter the protein structure and thus its function.

Finally, it is important to note that both mentioned studies, while not strictly clinical genomics works, were foundational in establishing the use of genomic analysis for determining the genetic causes of Mendelian diseases. Following the publication of these works, there was an explosion in the use of MST for the discovery of genes responsible for Mendelian diseases, leading to the identification of the nearly ten thousand genes responsible for them, as previously mentioned. However, we must emphasize that the importance of knowing the gene and the responsible mutations is that, as in the previous example, this knowledge enables accurate genetic diagnosis in other cases presenting similar symptoms. This goal, as we will see next, is one of the main objectives of clinical genomics.

The First Case of Diagnosis (and Treatment) by MST

The first published case of a molecular diagnosis made by MST appeared in the prestigious journal Science Translational Medicine in 2011. In this case, using whole genome sequencing, researchers identified the causative variants for a rare neuromuscular disease in twin siblings Alexis (female) and Noah (male) Beery from San Diego, California. The medical history of both began at the age of two when they were diagnosed with cerebral palsy after presenting various neuromotor difficulties. However, due to the typical variation in symptom severity throughout the day and other clues, the diagnosis was reevaluated; at five years old, they were diagnosed with L-dopa responsive dystonia (DRD).

DRD is a Mendelian disease, which can present with either a dominant or recessive inheritance pattern, and whose cause is the body's inability to produce the neurotransmitter dopamine. This inability is associated, in cases with a recessive inheritance pattern as appeared in the present case (both parents and an older brother are healthy), with the presence of two non-functional alleles for the gene encoding Tyrosine Hydroxylase (TH), the protein that converts the amino acid tyrosine into L-dopa, which is then transformed into dopamine by other neuronal proteins. The change in diagnosis was excellent news, as DRD is treatable with low doses of L-dopa, the same drug used to treat Parkinson's disease. The rapid reduction of symptoms after starting the L-dopa treatment seemed to confirm the diagnosis. However, some symptoms persisted and even began to worsen over time. At age 4, Noah experienced tremors and attention problems, while Alexis suffered from increasingly severe respiratory issues. Additionally, sequencing of the TH gene by the Sanger method did not reveal any known variants associated with DRD, thus increasing the concern for the health of both patients.

Through a contact of the patients' father, the case was evaluated by scientists and physicians at Baylor College of Medicine who suggested using genomics to search for an accurate diagnosis. Dr. Gibbs's team decided to determine the whole genome of both twins, their healthy older brother, and both parents, to try to determine which of the more than two million variants could be responsible for the pathology. Using filters similar to those described in previous works, looking for variants that could alter protein function, with low (or zero) frequency in healthy individuals, focusing on a select group of candidate genes (those associated with neuromuscular function) and analyzing familial segregation, they finally identified two pathogenic variants in the same gene. Remember that in a recessive model, what is looked for is the presence of two non-functional alleles that must be shared by both affected siblings, but not by the healthy brother, and since both parents are healthy, they must both be heterozygous carriers (possess one of the pathogenic alleles and one healthy allele). The study confirmed the absence of mutations in the TH gene and also in other genes associated with DRD, but enabled the identification in both twins of two potentially non-functional variants for a protein called Sepiapterin Reductase (SPR), responsible for manufacturing a compound called tetrahydrobiopterin. This compound is required not only by TH to make L-dopa but also by another protein responsible for making the serotonin precursor, 5-hydroxytryptophan. In short, the twins, due to the lack of SPR, were unable to produce both neurotransmitters, dopamine (associated with DRD) and serotonin. The mutant alleles corresponded, respectively, to a change from an A to a G, which leads to the replacement of the amino acid Arginine by Glycine, and an A to a T that results in the change of a coding codon for a STOP codon, resulting in a shorter and non-functional protein. The lack of SPR activity was biochemically confirmed in cell cultures from both patients, resulting in the definitive proof of the diagnosis.

An interesting point that arises from analyzing the present work is the analysis of the family tree. It spans four generations and shows that indeed there is no history of DRD in either family branch; however, both family branches show a recurrence of symptoms associated with neuromuscular issues. On the paternal side, women with depression have been reported in all generations, consistent with serotonin synthesis problems. On the maternal side in the current generation (second cousins of the twins) and their parents, unidentified neuromuscular symptoms are evident. This analysis of the family tree is a clear indication of the presence of both pathogenic variants run through the family lineage and confirms the suspicion of the inheritance model.

Finally, one of the most important points of the work and one of the direct consequences of the molecular diagnosis was that, immediately, it suggested a specific treatment. This treatment consisted of combining doses of L-dopa, the dopamine precursor, with 5-hydroxytryptophan, a serotonin precursor. With this combination of both precursors, the patients showed significant improvements in their health. Alexis, who had been struggling with severe respiratory problems, no longer exhibits these symptoms, allowing her, for example, to resume her sporting activities. Noah, for his part, began to show remarkable improvement in his overall coordination and attention capacity in class, which has had a positive impact on his academic performance and on his overall quality of life. This favorable response to the treatment has been a great relief to both patients and their families, underscoring the importance of an accurate diagnosis and an appropriate therapeutic approach. While this case may be considered fortunate in terms of suggesting an effective and straightforward treatment, it demonstrates the power of complete genome knowledge in achieving an accurate molecular diagnosis.

A Case Worthy of a Dr. House Episode

The second case I will tell you about is worthy of a Dr. House episode, although he, working in the pre-genomic era, probably would not have solved it. It was published in the journal Genetics in Medicine in 2011, and it shows the difficulty and effort required to find a diagnostic variant that would allow for progress toward treatment, and how the ingenuity of healthcare professionals enabled them to navigate the diagnostic odyssey to a successful outcome.

The case begins with the arrival of Nicholas Volker, a 4-year-old blue-eyed, Batman fan, who had arrived at the Children's Hospital of Wisconsin two years earlier with severe malnutrition. Nicholas's problem was that every time he ate something, his body generated an immense inflammatory reaction in his intestines (generically known as Inflammatory Bowel Disease, IBD), resulting in severe intestinal lesions, leading to the development of fistulas and bacterial infections due to bacteria escaping from the intestines into his internal environment. Nicholas had been a healthy baby until just before his second birthday. One day his mother discovered a small abscess at the exit of the rectum, took him to the hospital where they gave him antibiotics, but the abscess burst, and fistulas began to form. His pediatrician, Dr. Alan Meyer, had been treating the child with antibiotics and dozens of operations to heal his lesions and remove parts of his intestine damaged by ulcers. At the hospital, they had never seen a similar pathology, nor had colleagues consulted, or any searches performed in the available medical literature yielded clues for a possible cure. Nicholas's symptoms were similar to Crohn's disease as well as other IBDs. But these cases are usually treatable with corticosteroids for acute events, immunosuppressive drugs, and a strict diet. Nicholas's case did not behave in the same way, and genetic analyses of some genes for autoimmune diseases had come back negative. With this scenario, Dr. Meyer decided to try and to persuade his colleague, Howard Jacob, director of the Wisconsin Genetic Research Center, to use a new fancy and expensive technology, whole genome sequencing, to try and save a child's life.

Dr. Meyer wrote the following email to his colleague Dr. Howard in June 2009: "I am writing to inquire whether it is possible to sequence the complete genome of a pediatric patient, Nicholas, 4 years old. There is a high likelihood that the patient has a genetic defect and possibly a novel disease. A rapid diagnosis could save his life and truly demonstrate a case of personalized and genomic medicine." Although Howard was tempted to accept the proposal, he had some doubts. After researching the literature, he found no documented case of a patient diagnosed from their genome (recall that the first case of identification of a gene responsible for a known disease was recorded in 2010). Howard estimated the cost at about 75,000 US dollars, a figure that wouldn't be covered by health insurance. He had planned to apply genomic analysis in clinical practice, but not until 2014; Alan's request was five years early. Despite everything, he decided to accept the challenge.

As in the previous case, to search for candidate variants, the researchers focused on nonsense or missense variants that were not present in healthy patient databases and that were in genes associated with the immune system. Since the pathology was unknown, they had to consider both de novo dominant models, as well as recessive models with possible compound heterozygosity, and also X-linked models. The analysis identified a candidate variant located in the XIAP gene, located on the X chromosome and associated with immune system regulation. The variant was new, i.e., not reported in any database, and recent work connected XIAP indirectly with inflammatory bowel diseases. The variant involved a change from a G to a T, resulting in the replacement of a Cysteine with Tyrosine (recall the recurrence of Cys changes in pathogenic variants). With that level of evidence, the variant was interesting and potentially pathogenic, but researchers could not yet ensure it was responsible for Nicholas's phenotype. Therefore, they decided to take it as a working hypothesis and deepen the analysis of the variant.

First, they deepened the analysis of the variant's frequency. Since at that time there were not many genomes/exomes available, the researchers wrote to all their colleagues who they knew were working on exome/genome sequencing to check if the variant was present in any of their studied samples. After writing and receiving responses to many emails, they confirmed that among about 2000 exome/genome samples reviewed, none presented Nicholas's mutation. Additionally, a consultation with a bioinformatician friend who conducted a Multiple Sequence Alignment confirmed that in XIAP's homologous protein, cysteine is 100% conserved in chimpanzees, mice, cows, and even flies, indicating its functional importance. Finally, the definitive confirmation was provided by Dr. Verbsky, an immunologist who joined to collaborate on the case. Drawing on the available literature information about the XIAP gene, known to play a central role in the pro-inflammatory response and apoptosis, they reasoned that Nicholas's problem was that the malfunctioning XIAP resulted in a failure to control the immune system's response to bacteria naturally found in our digestive system. In other words, Nicholas's immune cells were hyperreactive. To confirm this idea, the researchers designed two functional assays to evaluate Nicholas's immune cells' response compared to those of a healthy donor. Both tests, the first based on macrophage stimulation and the second on apoptosis induction, showed that Nicholas's cells were hyperreactive and failed to turn off the immune reaction. These results, confirming the pathogenicity of the variant, also provided a potential therapy: replacing Nicholas's faulty immune system with one that functions correctly. The immune system replacement was performed through an "allogeneic hematopoietic precursor cell transplant." The success was demonstrated when, 42 days later, Nicholas was able to eat normally without presenting symptoms of recurrent Intestinal Inflammation, which finally led to his discharge and the start of a normal life. Once again, as in the case of the twins, Clinical Genomics had saved a life.

As shown by the previous examples, Fig. 6.1 visualizes how the introduction of massive sequencing technologies acts as the critical turning point in a patient's diagnostic odyssey, moving from years of uncertainty and inconclusive testing to a definitive molecular diagnosis that guides future targeted management."

The Importance of Patients with Mendelian Diseases for Understanding Gene Function

To conclude this chapter, I would like to offer a brief reflection. In his outstanding book Genome: The Autobiography of a Species in 23 Chapters, published in 2006, Matt Ridley reminds us, "Open any catalog of the human genome and you will face

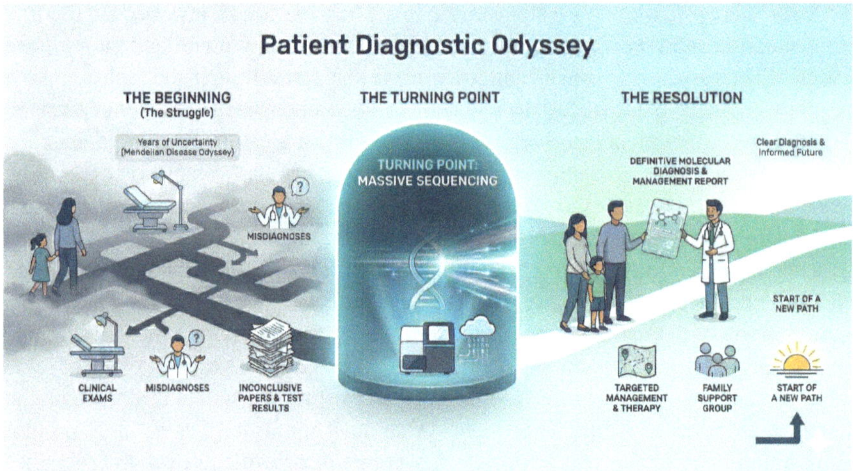

Fig. 6.1 The patient diagnostic odyssey (**A**) The Beginning (The Struggle). (**B**) The Turning Point: MASSIVE SEQUENCING TECHNOLOGIES. (**C**) The Resolution. A definitive molecular diagnosis.

not a list of human potentialities, but a list of diseases, most of them named after obscure pairs of Central European doctors." But why do we often name or describe many genes according to the diseases they cause when they fail, and not according to their physiological role in normal health conditions? This fact, which fortunately has significantly diminished after the Human Genome Project, reflects two important realities. On one hand, it highlights our historical ignorance about the true functions of these genes that persisted throughout the twentieth century, and on the other hand, it emphasizes that a patient homozygous for a deleterious allele in a given gene (let's call it gene X), serves as a natural version of what scientists would call a human knockout (KO) for gene X. The detailed analysis of these "natural knockouts" provides invaluable information about gene functions, as observing what fails in the absence of a particular gene helps deduce its normal role in human physiology. As we discussed in this chapter, the advancement of the Human Genome Project and MST have accelerated the study of more and more genes associated with Mendelian diseases. Combined with the study of these through molecular and cellular biology techniques, we achieve a more comprehensive understanding of the human genome, facilitating the development of targeted new therapies and personalized medicine strategies. The integration of all this knowledge has been crucial for transitioning from a vision focused on "a gene and its disease" to a holistic perspective of our health.

This chapter therefore provides us with one of the most relevant impacts of clinical genomics, which is the explosive exploration of genetic phenomena through the lens of Mendelian diseases, which further underscores the MST transformative approach. By dissecting the nuances of these diseases, scientists and clinicians aspire not only to enhance disease management and therapeutic interventions but also to fortify the fundamentals of genetic education and preventive medicine. Notably, this nexus of genomic insights and clinical applications illuminates the pathways to not just treating conditions but preempting them, heralding a future where comprehensive wellness is accessible and actionable.

Thus, as we harness these insights, let us recognize the patients and their unique genetic narratives, not as mere subjects of scientific inquiry but as central figures in the unfolding story of medical progress—a story punctuated by empathy, precision, and visionary foresight, culminating in a legacy of healthier future generations. While Mendelian diseases provided the first proof of concept for clinical genomics, the same principles are now being applied to the complex landscape of somatic mutations in cancer and the diversity of populations worldwide, as we will analyze in future chapters.

References and Suggested Further Readings

Ng SB, et al Targeted capture and massively parallel sequencing of 12 human exomes. Nature. 2009;461:272–6. https://doi.org/10.1038/nature08250

Ng SB, et al Exome sequencing identifies the cause of a Mendelian disorder. Nat Genet. 2010;42:30–5. https://doi.org/10.1038/ng.499

Bainbridge MN, et al Whole-genome sequencing for optimized patient management. Sci Transl Med. 2011;3(87):87re3. https://doi.org/10.1126/scitranslmed.3002243

Worthey EA, et al Making a definitive diagnosis: successful clinical application of whole exome sequencing in a child with intractable IBD. Genet Med 2011;13(3):255–62. https://doi.org/10.1097/GIM.0b013e3182088158

Boycott 1 KM, Vanstone MR, Bulman DE, MacKenzie AE. Rare-disease genetics in the era of next-generation sequencing: discovery to translation. Nat Rev Genet. 2013;14(10):681–91. https://doi.org/10.1038/nrg3555.

Gibson G, Muse SV. A primer of human genetics. Oxford University Press; 2020.

Chapter 7
Annotation and Classification of Variants According to the Criteria of the American College of Medical Genetics (ACMG)

Abstract This chapter presents the ACMG (American College of Medical Genetics) framework for the standardized annotation and classification of genetic variants, a cornerstone of clinical genomics. It traces the historical development of the guidelines, culminating in the 2015 ACMG criteria, which formalized a systematic, evidence-based approach to variant interpretation. Central to this framework is a two-step process: first, assigning evidence-based labels that reflect pathogenic or benign characteristics, and second, combining these labels according to defined rules to classify variants into five categories: pathogenic, likely pathogenic, variant of uncertain significance (VUS), likely benign, and benign. The chapter details the different levels of evidence, ranging from very strong to supporting, and highlights the diverse types of data used in classification, including population frequency, molecular effect, computational predictions, functional assays, and familial segregation. It further emphasizes the importance of integrating gene-specific knowledge and clinical context. Special attention is given to VUS, discussing their heterogeneity and the challenges they pose for diagnosis and decision-making. Finally, the chapter outlines strategies for efficient annotation and prioritization, reinforcing the ACMG system as a dynamic, interpretative framework that combines computational analysis with clinical expertise to support accurate genomic diagnosis.

Keywords ACMG criteria · Variant classification · Pathogenicity evidence · VUS · Clinical interpretation

The Origins

In the year 2000, in the journal Genetics in Medicine, the "ACMG Laboratory Practice Committee Working Group" published the first recommendations for the standardization of the interpretation and reporting of variants identified in the context of genetic

M. A. Marti, *Clinical Genomics*,
https://doi.org/10.1007/978-3-032-34667-4_7

studies with two main goals: first, to provide a unified framework for these types of studies; and second, to contribute to the education of physicians about the possible outcomes of genetic studies, so that they can adequately inform their patients and relatives. In this work, it was recommended to classify variants into five categories: (1) Variants previously reported and recognized as a cause of the pathology. (2) Variants previously not reported, but which according to their characteristics are expected to be causal of the pathology. (3) Variants not reported and that could, or might not be, causal of the pathology (pathogenic), (4) Variants not reported but that likely are not causal of the pathology, and (5) variants previously reported that are known not to be causal of the pathology, but are polymorphisms (or natural variants). Additional recommendations were included to aid the study of variants, such as the analysis of familial segregation and the development of functional studies. In 2008, also in Genetics in Medicine, these criteria were revised. The five previous categories were maintained, and a sixth was added) for variants that are not expected to be causal of the pathology but are variants associated with a clinical presentation. The focus was on using standardized terminology and nomenclature for the variants and progress was made in the discussion of the use of different tools. In this work, reference began to be made to variants as "Benign" and "Of Uncertain Significance" (that is, of uncertain meaning), although the term "deleterious" was preferred over "pathogenic" for the variants causing disease. These recommendations saw the birth and exponential growth of MST in their applications in clinical genomics, and particularly in the study and diagnosis of Mendelian diseases as seen in the previous chapter. As a result of this explosion in the use of technologies, in 2015 the ACMG published an updated set of criteria, which not only indicate how to classify variants in the known categories (Pathogenic, Variant of Uncertain Significance or Benign) but also how to annotate and assess the annotation of the variants. These criteria are what we now call the ACMG criteria.

The 2015 ACMG Criteria for the Assessment and Classification of Variants

With the aim of standardizing the growing number of variant reports based on the use of MST, and recognizing that the previous recommendations did not provide a detailed algorithm that allowed for their classification, in 2013 a working group was formed consisting of physicians and laboratory professionals to develop a method of variant classification based on the critical and systematic evaluation of the evidence presented by each variant. The result is a classification system with five tiers, corresponding to the known (1) Pathogenic (P), (2) Likely Pathogenic (LP), (3) Variant of uncertain significance (VUS), (4) Likely Benign (LB), and (5) Benign (B). But how is it determined to which category each variant belongs?

To classify the variants, we will start with their annotation. To organize this, the recommendations define six levels of evidence (or tags) that evaluate 28 different characteristics that a variant may, or may not have. The six levels of evidence, in decreasing strength of pathogenicity are PVS (Pathogenic Very Strong), PS (Pathogenic Strong), PM (Pathogenic Moderate), PP (Pathogenic Supporting), BA (Benign Stand Alone), BS (Benign Strong), and BP (Benign Supporting), and for some of them, different characteristics are evaluated which results in more than one tag with the same strength, for example, PM1, PM2, PM3, etc. As we will see below, each tag (or level of evidence) entails a specific characteristic of the variant, such as being "de-novo," which must be evaluated and, in the event the variant possesses the characteristic, it is "assigned" (or placed) the corresponding tag instead. However, if the variant does NOT possess the characteristic, the tag is not placed. The result of this process, the tagging, (which is nothing other than an enhanced annotation), is that each variant will have certain tags. For example, a variant may have PS2, PM2, and PP3 while another may have such tags as BA and BP3. For each variant, the combination of its tags results, according to strict combination rules, in its classification into one of the five categories.

In summary, the classification of variants according to ACMG-2015 is based on a two-step process. The first involves tagging each variant, and the second combines the tags to classify the corresponding variant. Let's now look at the tags one by one.

Characteristics, Labels and Levels of Evidence with Pathogenic Traits

PVS1. This label is assigned to those high-impact variants (nonsense, frameshift, canonical splicing site variants, or deletions of one, or more, exons) in a gene where loss of function (LOF), is a known mechanism of pathological development. As we will see with several labels, PVS1 evaluates two aspects. First, the nature or type of the variant, and second, the characteristics of the gene containing it. The first characteristic is completely objective, as once the variant is defined, there is no doubt about its molecular effect. The second, associated with the gene, has a degree of subjectivity since we must judge whether the gene is a gene where LOF is a known mechanism. Generally, these genes are those that have extensive knowledge of their molecular function, and how its failure leads to disease. Paradigmatic examples of these are the causal genes of recessive diseases, such as phenylketonuria or cystic fibrosis.

PS1: is assigned to those variants that produce the same amino acid change as a different variant previously established as pathogenic. This label results from the degeneracy of the genetic code. For example, the TTT codon encodes Phe. Suppose the T>A variant in the third position, which results in the TTA codon that encodes for Leu, was previously reported as pathogenic. If the variant resulting in a T>C change in the first position is now present, it results in the TCC codon, which also encodes for Leu. In this case, this variant is assigned PS1. Although this label seems objective, it is important to consider that the criterion to decide if the reference variant is pathogenic or not has a certain degree of subjectivity.

PS2: Is assigned to those variants that have been confirmed to be "de novo," having confirmed the paternity of both parents and in the absence of symptoms in other family members. Clearly, to assign this label, it is necessary to have the exomes/genomes of both parents, and to verify both the absence of the same in both as well as their paternity, and/or have performed a confirmation of the variant by Sanger sequencing and an independent paternity confirmation.

PS3: Is assigned when in-vitro or in-vivo assays, well established, support the deleterious effect of the variant on the gene and/or its product (the protein). This label, one of the most difficult to place and most subjective, requires in a first approximation, some scientific work where an in-vitro assay, such as cell culture and/or enzymatic/ protein activity assays, and/or in-vivo, such as a transgenic mouse, has specifically evaluated the variant. It also usually requires that the same assay presents results from negative controls, with known benign variants, and especially from positive controls, with previously reported pathogenic variants.

PS4: Is assigned when the prevalence of the variant (i.e., its frequency) in affected individuals is higher than the prevalence in control individuals. Put another way, it is assigned when a low-frequency variant is observed in multiple unrelated affected individuals but is absent in healthy individuals. It is interesting to note, that this label has two relevant degrees of subjectivity: on one hand, what is considered as low frequency. Usually, the threshold should be below 1 in 1000 or even 1 in 10000. On the other hand, it must be observed in multiple affected individuals. But how many are multiple? A loosen criterion would be at least two, although in practice "multiple" is more often associated with numbers greater than three, or even more.

PM1: This label is assigned when the variant is of the missense type and is located in a well-established functional domain of the protein, such as the active site. Furthermore, it must be shown that all missense variants identified in the domain are pathogenic, and the domain must not have any benign variants reported. The key in this label is defining the well-established functional domain, since this allows, or prevents, the label to be assigned to the variant. If we use a very strict definition of functional domain, for example, in an enzyme the active site residues that are essential for carrying out the reaction, such as the catalytic triad Ser-His-Asp of serine proteases. Then, if we have a variant that for instance results in the change from Ser to Ala in the first of them, it is clear that the variant is in the functional domain. Noteworthy, since the three residues are essential, by the pure definition of essential), any change in these will result in an inactive protein, and hence the corresponding variants will be pathogenic and there will be no benign variants. If a looser criterion is used to define the well-established functional domain, subjectivity increases.

PM2: This label is primarily associated with the allelic frequency of the variant. It is assigned when the variant is absent in control individuals, thus it is not reported in databases of healthy individuals, or in cases of recessive diseases, the frequency is less than expected. Therefore, to analyze whether or not to place this label, we must first look at the frequency of the variant. If the variant does not have a frequency, which implies it has never been observed, then the label is straightforwardly placed. If the variant has a frequency, it is necessary to see if the gene that contains it is known to cause a recessive disorder, and in that case, even if the variant has been observed in a certain number of individuals, and therefore a population frequency of the variant can be calculated, the label can be placed only if the frequency is very low (less than expected). The subjectivity of this label lies in that one must define what the expected frequency is that sets the threshold for placing or not placing the label, and in knowing the disease model (dominant or recessive). In practice, the usual way to place it is to use a rather low threshold, as mentioned in PS4, in the order of 1 in 10,000. Moreover, an even more advanced way to place it, as we will see below, defines a frequency threshold that is specific for each gene.

PM3: This label is placed when it is positively known that the gene containing the variant causes a recessive pathology, and another pathogenic variant in trans has already been found in the same patient. In other words, the label is placed on a second variant that can lead to a compound heterozygote when a pathogenic variant has already been found in the same patient, and a segregation study has been conducted to confirm that the variants are in trans, in other words, they are in the affected patient each on a different allele. If working with exomes, the study to verify the variant is in trans requires having the exome/Sanger of both parents to confirm that each of them is heterozygous for one of the variants. Alternatively, if one has the complete genome, although complex, sometimes it is possible to verify in the patient's sample alone that both variants are on different alleles.

PM4: Is placed when the variant comprises a small insertion or deletion (indel) that does not alter the reading frame, thus resulting in a change in the length of the protein. In addition, the indel must be located in a non-repetitive region of the protein. The first part of the label is easy to verify and objective, since it requires that the indel comprises a multiple of three bases. To verify that it is not in a repetitive area is also easily verifiable because these are usually annotated. The origin of this restriction for the label arises from the observation that a significant number of proteins have repetitive zones, such as polyglutamines, and in these, a small change in the number of repetitions generally does not have a pathogenic effect, as it requires a significant increase in the repetitions. For example, in the gene causing Huntington's disease (HTT), the gene has a trinucleotide repeat region "CAG" that results in a glutamine (Q) repeat in the protein. Benign alleles usually have fewer than 26 repeats with a wide variability. In this case, an indel that eliminates/adds between 1 and 10 repeats is not pathogenic, as only above 36 repeats does the risk of developing Huntington's disease increase.

PM5: This label is placed when the variant is of the missense type and in the same position in the protein, there is another different missense variant that is pathogenic. For example, if the variant we are analyzing is the Cys44Ala change, and it is known that Cys44Ser for instance is pathogenic, then PM5 will be assigned to Cys44Ala.

PM6: Is similar to PS2, but when paternity cannot be confirmed. It is placed if the variant is assumed to be de novo, but without paternity confirmation. When we have a variant observed in the affected individual and that is absent in the parents, we will place PM6 or PS2. If we only have the Sanger from the parents, and do not have an independent assay to verify paternity, PM6 will be placed; however, if we have the exome/genome and can verify paternity, the PS2 label will be placed.

PP1: This label is placed when the variant co-segregates with the disease in multiple family members, and also when it is known that the gene containing it, is associated with the same disease. In other words, to place this variant label, we must know that the gene containing the variant is associated with the disease and be able to analyze its presence, for example, by target sequencing using the Sanger method, in several affected family members. This implies going beyond the typical mother-father-affected child trio, usually verifying the presence in siblings, cousins, or affected uncles and aunts.

PP2: Is placed on missense type of variants that are found in genes where "the majority" of the missense type variants are (Likely) pathogenic. This label, as we have already seen with others, has the problem of not precisely defining what "the majority" means. For instance, we could consider the majority as "more than half" (the famous simple majority), or if we want to be more strict, we might require that more than two-thirds of the variants be pathogenic. The problem with these simple metrics is that with the number of variants detected today in massive genome projects, many variants are known for almost all genes, most of them of the VUS type. Therefore, if we follow the rule strictly we would never place the PP2 label. On the other hand, it is different for a gene in which three pathogenic variants out of a total of 4 are known, compared to a gene in which 30 pathogenic variants out of 100 are known. Therefore, for this label to be "useful" and be placed with confidence, we must properly evaluate the gene as well as the variant. What the criterion "looks for" is to place PP2 on those genes where the missense variants tend to be pathogenic, and this happens when the gene is not tolerant to mutations. In other words, PP2 is placed when we find a missense variant in a gene that is intolerant to this type of mutations. For this, as we will see later, we must evaluate the genes according to the available information derived from the set of known variants found in the gene.

PP3: Is placed when "multiple" computational models support that the variant has a deleterious effect on the gene or the protein. In particular, it is suggested to evaluate models based on information about conservation, evolutionary, structural, splicing, etc. In other words, we place this label when "multiple" variant effect prediction programs suggest that it is pathogenic. This label, while it might seem objective, since it is based on the result of a prediction algorithm, is highly subjective since it requires deciding which program(s) to use, what confidence threshold in relation to the pathogenicity criterion to use in these programs, and especially how many programs represent the multiple criteria. In practice, it is suggested to use between two to five different predictors, provided that they are based on different biological principles. If the most recent predictors are used, such as meta-predictors that already combine various properties and even several predictors, then it is even possible to use just one.

PP4: This label is placed when the patient's phenotype (or their family history) is consistent with that expected for the causative gene, and this gene is associated with a specific disease. In other words, this label is placed when we find a variant in a gene that is expected to be causative gene, given the available clinical picture (or presumptive diagnosis) observed in the patient.

PP5: Finally, this last label associated with the pathogenicity of the variant must be placed when the variant we are evaluating has already been previously reported as pathogenic. Note that, according to the recommendations, the fact that the same variant we are evaluating has been reported as pathogenic does not result in its classification as pathogenic per se, but rather it is an additional label that contributes to its pathogenicity. To classify it as such, however, we must perform the labeling, and complete classification ourselves.

Having finished the analysis of the labels with pathogenic characteristics, we will now analyze those with benign characteristics.

Characteristics, Labels, or Levels of Evidence with Benign Traits

Benign character labels are significantly fewer than those of pathogenic nature, usually easier to assign and, as we will see, many are the other side of the coin of a pathogenic label with the same force. But let's start with the single and most important label BA1.

BA1: This label, called "Benign Stand Alone," is the only label that by itself is sufficient to classify a variant as benign. A variant with this label will consequently be classified as benign. It is assigned if the frequency of the variant is greater than 5%. This criterion, which, given the strict definition of the threshold, is completely objective, is based on the fact that we are evaluating variants that are potentially causal of rare diseases. If a causal variant of a rare disease had a frequency above 5%, more than 2.5% of the population would have the pathology, and therefore it would not be rare. From a biological perspective, and as we have already mentioned in other chapters, if a variant is pathogenic, it results in a disadvantage for its carrier, and therefore, the process of natural selection would tend to eliminate it from the population, resulting necessarily in its observed low frequency.

BS1: Follows the same logic as BA1, and is placed when the frequency of the variant is higher than expected. As we can see, in this case, unlike the previous one, the threshold is not fixed, and will depend on the gene. In other words, what we need to ask ourselves is, what is the minimum frequency we can expect for a pathogenic variant in this gene? What it seeks to evaluate for the gene threshold is what would be the expected frequency given the observed prevalence of the disease. In diseases with very low prevalence, for example, 1 in 100,000 individuals, then if the frequency of the variant is more than 1%, we will place BS1, which strongly suggests that the variant is benign. Another point to consider regarding BS1 is that it is the opposite label to PM2. That is, given a variant according to its frequency and the expected threshold for pathogenic variants in that gene, it will either be assigned PM2 or BS1.

BS2: Is placed when the variant has been observed, with the appropriate zygosity in a healthy individual. This label is related to the penetrance of variants, or as mentioned in previous chapters, with the "magnitude of the effect" (Odds ratio). As we have previously described, we assume that pathogenic variants have an infinite Odds ratio, which is equivalent to saying that they have 100% penetrance, or in simple terms, that if one has the variant (or two in the case of a recessive disease), one will definitively develop the pathology. However, since there are variants and diseases with incomplete penetrance, it is possible for a pathogenic variant to be found in a healthy individual, and if this is observed, BS2 will be placed on the variant and the other labels must be evaluated to determine if the variant is ultimately confirmed as benign.

BS3: Is the opposite label to PS3, and is placed when in a well-established functional study, it has been shown that the variant does not significantly alter the function of the gene or protein. Therefore, the same recommendations and precautions mentioned for PS3 apply to this label.

BS4: Is the opposite label to PS4 (or with less force, PP1) and is placed when the variant does not segregate in the affected family members. As an example, this label would be placed if we have a variant that is present in the patient, but, for example, is not observed in an affected sibling or cousin.

BP1: This label is placed when what we have is a missense type variant, but the gene containing it primarily has known pathogenic variants of the nonsense type. This label is directly related to PP2, where we must analyze the gene's capacity to tolerate missense type variants or the lack thereof. If the gene is considered tolerant to missense type variants, since the nonsense type will primarily be pathogenic, BP1 will be placed.

BP2: This label is placed when there is another variant that already explains the case. In other words, the variant to be labeled is found in the same gene and in trans with a variant already classified as pathogenic in the case of a dominant disease, or is observed in cis (on the same allele) with another variant previously classified as pathogenic for both dominant and recessive inheritance patterns.

BP3: This is the equivalent case to PM4 but when the indel falls in a repetitive area like the case mentioned for the gene associated with Huntington's disease. In other words, if we have an indel that does not alter the reading frame but is in a repetitive zone, we will place the label.

BP4: Is the opposite label to PP3 and is placed if "multiple" computational evidences suggest that the variant is benign.

BP5: Is placed when there is an alternative explanation for the pathology in terms of variants. This label is similar to BP2, but in this case what we have, is not another variant in the same gene, but another variant classified as pathogenic in another gene whose expected phenotype is consistent with that observed in the patient.

BP6: Is the opposite label to PP5, and is placed when the same variant we are analyzing has been previously classified as benign.

BP7: Finally, this label has some overlap with BP4, as it is placed when dealing with a synonymous variant, in which splicing algorithms do not predict a significant effect on this process, and when the variant is also in a less conserved position.

Having analyzed each label individually, before conducting an in-depth joint analysis of them, and outlining strategies for their efficient implementation, let's briefly examine how labels combine to determine the final classification according to the ACMG criteria. And remember that these criteria are not static; they require the integration of computational data, population frequency, and clinical intuition.

Classification of Variants According to ACMG Based on Labels

According to the ACMG's 2015 work, once labels are assigned, "combination rules" must be used to determine the classification of each variant based on its tags. Let's analyze what combination of labels results in each of the five categories. For a variant to be classified as pathogenic, it must possess the following combination(s) of labels: one PVS label and a second PS label (or alternatively, two or more PM or PP labels) Alternatively, a variant will also be classified as pathogenic with two or more PS labels, or one PS and three PM labels (or two PM and two PP, or one PM and four PP) In other words, for a variant to be pathogenic, we need two or more (very) strongly pathogenic labels, and in case of having only one, we can replace it with two to three moderate labels, or even more supporting labels.

As an example of a pathogenic variant, consider: NM_000277.2(PAH):c.1A>G (p.Met1Val). The PAH gene encodes for Phenylalanine hydroxylase, responsible for hydroxylating phenylalanine and converting it into tyrosine, one of the first steps in various metabolic pathways, and whose loss of function in both alleles leads to phenylketonuria (PKU) with an autosomal recessive inheritance model. The variant has a population frequency of 0.00002, so PM2 is assigned. The variant has been studied in an in-vitro assay of enzymatic activity presenting less than 3% activity compared to the wild protein, resulting in the assignment of the PS3 label. The variant has been observed in trans with known and clearly pathological variants as in numerous cases in patients with PKU, therefore, it is labeled PM3. The gene where the variant is located is very specific for PKU phenotype observed recurrently in patients with this variant or labeled with PP4. Additionally, the variant has been previously reported as pathogenic, so it is labeled with PP5. According to ACMG criteria 1PS + 2PM + 2PP, the variant is then classified as pathogenic.

Now analyzing the combinations of labels that lead to Likely Pathogenic variants, we see that it requires: A PVS with a PM label (for pathogenic it was PVS + PS), or one PS and one PM (for P it was PS), or one PS and two PP labels. Alternatively, a variant can be classified as likely pathogenic with three PM, or two PM and two PP labels, or only one PM and four PP labels. This shows that for a variant to be likely pathogenic, it requires at least a moderate level label, and in that case, it must be complemented with 4 supporting types.

As an example of a likely pathogenic variant, we have in the same gene as the previous case the variant NM_000277.2(PAH):c.581T>C (p.Leu194Pro). The variant is absent from databases like 1000 Genomes and has a very low frequency in gnomAD (MAF: 0.00004), so it is tagged with PM2. The gene is specific to the pathology, so it is tagged with PP4. Like the other variant, this one has also been observed in trans with other pathogenic variants, particularly one that creates a stop codon, hence it is assigned PM3. The variant is predicted as pathogenic by several predictors, thus it is also tagged with PP3. In summary, with 2MP and 2PP, the variant is classified as likely pathogenic. It is interesting to compare both variants used as examples and observe that they have the same labels, but the first one, classified as pathogenic, is due to the existence of functional assays that allow assigning a strong criterion of pathogenicity (PS3).

Now let's turn our attention to benign variants. As already mentioned, a variant is considered benign if it has: The BA1 label alone, or alternatively two BS labels. In this case, it is worth highlighting that if the variant does not have a frequency high enough to be considered benign per se (BA1), in addition to the frequency, either the existence of another variant that explains the case (BS2), a functional assay (BS3), and/or incorrect segregation in the family (BS4) is required.

As an example of a benign variant, in the same gene, we have the variant: NM_000277.2(PAH):c.1242C>T (p.Tyr414=). This variant is synonymous, and we can assign the following two labels. On the one hand, it has been observed in 19 healthy homozygous individuals, which is not compatible with an autosomal recessive disease, so it is labeled with BS2. Additionally, its frequency MAF = 0.01361, which is higher than expected, so it is labeled BS1. Having two strong benign labels, it is classified as Benign.

For a variant to be likely benign, the requirements are clearly less stringent since it requires: one BS label and one BP, or alternatively two BP labels. As we see, the minimum requirement for a variant to be likely benign is two labels, unlike supporting level evidence such as computational evidence, being a missense in a gene that tolerates this type of mutation, an indel in a repetitive area, the variant having already been reported as pathogenic or the existence of an alternative explanation for the case.

As an example of a benign variant, we have NM_000277.1:c.772C>T12. Again, a synonymous variant. Similar to the previous one, this variant is assigned BS1 because its frequency is higher than expected. On the other hand, it is labeled with BP7 because it has a deleterious bioinformatics prediction. As a result (1BS and 1 BP), the variant is classified as likely benign. The difference with the previous variant is that I do not have the second BS that provides evidence from observing healthy homozygous individuals.

Finally, for a variant to be classified as a Variant of Uncertain Significance (VUS), there are two possibilities, which, as we will see later, require different analysis and treatment. In the first case, those variants that do not have enough labels to be classified as Likely Pathogenic (with at least 1 PM and 4 PP) or likely benign (at least 2 BP) will be reported as VUS. These are the most frequent VUS, since they are the ones for which very few labels could be placed. The second case, where a variant is classified as VUS, occurs when it possesses labels of both types, labels with pathogenic characteristics and labels with benign characteristics. Note that, according to ACMG 2015, a minor contradiction, even the presence of "a" single benign criterion in a variant with several pathogenic criteria, is enough for it to be classified as VUS.

As an example of a VUS variant, we have NM_000277.2(PAH):c.169G>A (p.Glu-57Lys). The variant has a low frequency, so it is labeled PM2. The variant is in a patient diagnosed with PKU, so it adds PP4. Since the variant has not been observed in trans with a known variant, there are no functional assays, and the bioinformatics prediction is not conclusive, no more labels can be assigned, so with only 1PM and 1PP, the variant does not have enough evidence to be classified as likely pathogenic, and therefore is a clear VUS.

Thus to summarize, the ACMG protocol for classifying variants consists of: First evaluating each label for each variant and deciding whether to apply it or not, and secondly based on the labels assigned to each variant to determine its classification. It is clear that this process, for the thousands of variants of an exome, or millions of a genome, is not performed exhaustively. That is, it is not performed by analyzing the 28 labels for each variant one by one, but by following different prioritization strategies that will allow efficient labeling and classification. To understand these strategies and the foundations of their operation, we will now proceed to analyze the labels not one by one, but together with a slightly more holistic view.

Analysis of ACMG Labels and Strategies for Efficient Placement

To understand in greater depth how to design a strategy for placing ACMG labels, the first step is to recognize that they can be grouped according to the biological concept or process being evaluated.

The first group we will analyze includes the assessment of the population frequency of the variant. Firstly, it is easy to see that all variants will have a population frequency value, either because they have been found in the databases of exomes/genomes, in which case the variant will have a specific value, or, if it has not been previously observed, we can assign a value of zero frequency. The labels associated with the frequency of the variant are mainly PM2 and PS4 associated with possible pathogenicity, and BS1, BS2, and BA1 associated with a benign character. Thus, if the frequency of the variant is higher than 5%, the variant will be assigned a BA1, indicating that it is clearly benign. If the frequency is higher than expected for recessive diseases, it will be assigned BS1, while its presence (frequency greater than zero) in healthy individuals will result in BS2. Alternatively, the absence (or very low frequency) of it in databases allows for the assignment of PM2, and its presence in affected individuals from different families allows for the assignment of PS4. This short analysis shows that frequency, as the first objective data of annotation, allows variants to be divided into two groups: those that are already on their way to being benign, and those that, if they continue to gather evidence, could be categorized as (Likely) Pathogenic.

The second group of evidence encompasses annotations and/or predictions of a variants molecular effects. These are related to the type of variant (missense, nonsense, silent), the location of the variant in the protein, and the knowledge available about the gene. These are PVS1, PS3, PM1, PM4, PP2, PP3, BS3, BP1, BP3, BP4, and BP7. If we analyze the nature of the variant, we can see that if it corresponds to a missense type, and it is known that loss of function of the gene is a known mechanism of pathogenic development, PVS1 will be placed. Conversely, if the variant is silent and does not affect splicing, BP7 will be placed. If the variant is an in-frame indel, we must analyze whether it falls in a repetitive site (BP3) or not in the protein (PM4). For missense variants, we must first analyze whether they fall at a functional site (PM1). Secondly, whether the gene is tolerant (BP1), or not (PP2) to mutations of this type. Thirdly, if there are functional assays, their results will allow the assignment of PS3 or BS3. Finally, bioinformatics prediction will allow the assignment of PP3 or BP4.

The third group of evidence depends on our knowledge of other variants in the same gene. This includes the labels PS1, PM5, PM3, PP5, BP2, and BP6. For any variant, previous reports of the "same" variant as pathogenic or benign will allow the assignment of PP5 or BP6. On the other hand, the presence of a known pathogenic variant in the same gene in trans and for a recessive pathology will allow the assignment of PM3, seeking to complete a compound heterozygote. While the presence of a known pathogenic variant in the same gene that can by its own explain the observed phenotype will allow the assignment of BP2. In the particular case of missense variants, and in relation also with molecular annotation, if there is a reported pathogenic variant at the position of the variant, PM5 will be assigned, and if the reported variant corresponds to the same amino acid change, PS1 will be assigned.

The fourth group of evidence depends on the particulars of the case and its family. This includes the labels PS2, PM2, PP1, PP4, BP4, and BP5. Firstly, we must analyze the segregation of the variant in the family. If it segregates correctly (meaning the affected individuals possess the variant with the correct zygosity, and the healthy family members do not possess it, or are healthy carriers), PP1 will be assigned. If segregation is not appropriate, BP4 will be assigned. Particularly if the analysis of segregation in the parents shows that the variant is de novo, and genetic paternity can be confirmed, PS2 will be assigned, and if it cannot be confirmed, PM2 will be placed. Finally, we must assess whether the phenotype of the patient or the family history is highly specific for those previously observed in the gene where the variant is found, in which case PP4 will be assigned. On the contrary if an alternative triggering cause of the observed phenotype is observed, which translates into the presence of another variant, perhaps in another gene with a higher level of evidence to explain the clinical picture observed, BP5 will be assigned.

Analyzing variants in a grouped way, results in a much more efficient manner of analysis and therefore allows for the design, as we will see later, of an efficient strategy for their prioritization in the context of their labeling and classification by ACMG. Another point to highlight from the group analysis is noting what type of information must be annotated on the variant to be able to carry out the labeling. For the first group, we need to know the allelic frequency of the variant, which is currently obtained directly from population databases. For the second group, we need structural annotation that tells us the nature of the variant and in which gene it is found. However, to be able to analyze all the labels, we will need to have information about the gene, mainly in relation to its tolerance to missense type variants, and about the protein in relation to the definition of functional sites and repetitive zones. To complete this annotation, we must also have bioinformatics predictors, and, if available, annotations associated with the existence of functional assays. The third group of labels, in addition to structural annotation, requires reliable information about the classification of other variants in the same gene. It is important to note that these three groups of variants depend only on knowledge of the variant and are independent of the case, so, as we will also see later, automatic annotation is possible. The fourth group, however, necessarily depends on knowing the details of the specific case and particularly on analyzing the variant in the context of the family, making automation difficult, if not impossible.

Deepening our Understanding of the Variants of Uncertain Significance (VUS)

Let's now delve into how the labeling of a variant relates to its classification as a VUS. First, let's recall what ACMG says about when a variant is classified in this category. Remember that there are two cases, which as we will see below, are significantly different. The first case comprises those variants whose level of evidence (the sum of their labels) is not sufficient for the variant to be classified as "Likely Benign," or on the other extreme, as "Likely Pathogenic." From the benign side, we see that the minimum number of labels necessary to be classified as Likely Benign is two BP labels. If a variant possesses at least one BP label (at the very least the one related to population frequency which can always be "placed"), then surely a thorough review will show that the other label, either BP or BS, will appear sooner or later. For example, it may be possible that bioinformatics prediction and/or knowledge of the gene will allow the missing label to be assigned to classify it as Potentially Benign. On the other hand, if any of these criteria result in potential pathogenicity, the variant will move to a VUS of the next category.

For those VUS that possess pathogenic labels, to reach a classification of likely pathogenic, it requires at a minimum one strong label (PS) and a moderate one (PM), or if not a strong one, two moderate or one moderate and four supporting (PP) labels. On this side of the mountain, many VUS appear that possess insufficient labels to reach the pathogenic classification. However, it is possible to see, according to the labels they do possess, not all these VUS will be the same. A large group of these will have very few labels, at least one associated with frequency (PM2), and perhaps a bioinformatics prediction. However, if little is known about the gene, labels associated with molecular effects cannot be assigned. If, additionally, there is no clear relationship between the pathology and the observed phenotype and/or the segregation in the family group is not known, it is likely that these variants will end up being benign and therefore there is little interest to deepen their study.

On the other extreme, there will be variants that possess several labels, for example, one PS or PM of frequency, and two to three PPs associated with the molecular effect. These labels are on the verge of being likely pathogenic. It is possible that they are also found in a gene for which the phenotype is well known and coincides with that observed in the patient. In this case, the VUS is a potential candidate for diagnosis but still falls short with the available evidence. Generally, what usually happens in these cases is that information about segregation in the family is not often available (which would allow placing an additional label confirming its pathogenicity). Or the current knowledge of the gene/protein is insufficient, and further studies are required, for example through functional assays. In these cases, it is recommended to advance in both directions, on one hand verifying the correct segregation of the candidate variant in the family group, and on the other, searching in the literature for research groups working on the gene and/or pathology that may be interested in including the variant in future functional trials. These variants are often reclassified as pathogenic (or sometimes as benign) over time due to the publication of functional studies and/or their appearance and reporting in other (unrelated) cases in the literature and/or variant databases. Therefore, for this typically small group of diagnostic candidate variants classified as VUS, it is suggested to perform a review of the labeling and classification periodically.

Moving on to the Second Group of VUS According to ACMG Criteria. These are, those variants that have "contradictory" labels, thus labels with pathogenic traits (PS, PM, or PP) and benign ones (BS, BP). However, the recommendations do not specify how both types of labels should be balanced. A literal and strict interpretation might suggest that any variant that has a contradiction should be classified as a VUS. But if, for example, a variant possesses the BA1 label (i.e., a frequency greater than 5%) and a pathogenic bioinformatics prediction PP4, a strict interpretation would imply classifying it as a VUS, which seems clearly erroneous. Even if a variant has 2 BS labels, for instance, a frequency greater than expected and a functional assay that showed no effect on protein function, a BP type label would force it to be classified as a VUS. On the other hand, we could have a variant with several pathogenic labels that would normally classify it as likely pathogenic, but due to the presence of a BP type label, it strictly should be classified as a VUS.

To advance in these cases, we must first analyze which label(s) are generating the contradiction and see if we can resolve it. For instance, for a variant that predominantly has benign labels, what is the pathogenic label? If it is a bioinformatics prediction, it should be examined in detail; if it is due to low frequency, perhaps the threshold is too high; if it is in a functional zone of the protein, analyze in more detail which residue change it corresponds to and what the definition of the functional site is, etc. In other words, it must be carefully reviewed whether the pathogenic label is well placed and how much confidence we have in it overall. On the other hand, if a variant mainly has pathogenic labels and only one benign one, it is necessary to see what it is and what its level of confidence is. In the case of frequency, for dominant pathologies the expected frequency is zero, but with nearly millions of exomes/genomes sequenced, it is possible that one to three alleles may be present, perhaps due to an error in the database. It is also possible that the bioinformatics prediction is wrong or that the knowledge regarding the gene/protein, its variants, functional residues, etc., needs to be reviewed. Beyond reviewing the contradictory label, a strategy for pursuing the potential reclassification of the variant is to ask the following question: what the classification would be if the contradictory label and an equivalent but opposite character characteristic label are removed?. For example, if a variant has 1 PS, 1 PM, 2 PP, and 1 BP. The BP label contradicts the others and suggests its classification as VUS. But if we eliminate BP along with one of the PPs, the resulting score would be 1 PS, 1 PM, and 1 PP, which is enough to classify it as likely pathogenic, suggesting a classification based on this model. In summary, within the group of VUS, we find variants with different characteristics according to the labels assigned to them; therefore, we must adopt a different strategy for their analysis and potential (re)classification.

As a concise summary, the intricate ACMG/AMP variant interpretation guidelines are shown as a comprehensive decision-support wheel in Fig. 7.1.

Having thus deeply analyzed the 2015 ACMG recommendations and how they frame, or practically define, the correct annotation of variants that results in their classification, in the next chapter we will analyze how to apply this framework in the context of clinical genomics, specifically when analyzing cases for the molecular diagnosis of rare diseases.

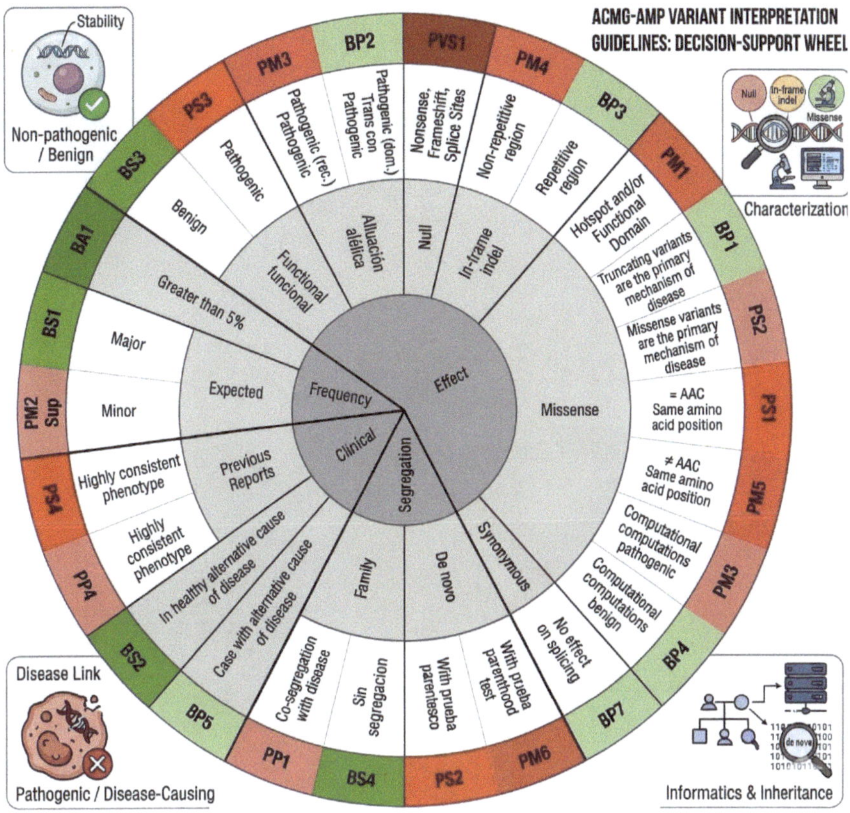

Fig. 7.1 ACMG-AMP variant interpretation guidelines: decision-support wheel. Green segments represent benign criteria (e.g., BA1, BS1) and red/pink segments represent pathogenic criteria (e.g., PVS1, PS1, PM1)

References and Suggested Further Readings

Kazazian H, et al ACMG recommendations for standards for interpretation of sequence variations. Genet Med. 2000;2(5):302–3. https://doi.org/10.1097/00125817-200009000-00009.

Richards CS, et al ACMG recommendations for standards for interpretation and reporting of sequence variations: revisions 2007. Genet Med. 2008;10(4):294–300. https://doi.org/10.1097/GIM.0b013e31816b5cae.

Richards S, et al Standards and guidelines for the interpretation of sequence variants: a joint consensus recommendation of the American College of Medical Genetics and Genomics and the Association for Molecular Pathology. Genet Med. 2015;17(5):405–24. https://doi.org/10.1038/gim.2015.30.

Oza AM, et al Expert specification of the ACMG/AMP variant interpretation guidelines. Hum Mutat 2018;39:790–805. https://doi.org/10.1002/humu.23428

Karczewski KJ, et al The mutational constraint spectrum quantified from variation in 141,456 humans. Nature. 2020;581(7809):434–43. https://doi.org/10.1038/s41586-020-2308-7.

Chapter 8
From Variant Analysis to the Report of Results

Abstract This chapter focuses on the transition from variant analysis to the generation of a clinically meaningful report in the context of rare disease diagnosis. It reframes the objective of clinical genomics from identifying strictly diagnostic variants to reporting those with broader clinical relevance, including pathogenic, likely pathogenic, and select variants of uncertain significance that may constitute diagnostic hypotheses. The chapter outlines criteria for determining which variants should be reported and emphasizes the importance of excluding those without clear clinical value to avoid ambiguity in medical decision-making. A structured strategy for variant prioritization is presented, beginning with phenotype-driven gene panels and progressing through systematic evaluation of variant classification, phenotype concordance, and technical validation. The workflow also addresses how to proceed in negative cases, including the verification of known pathogenic loci and the careful assessment of high-ranking VUS. Finally, the chapter details the essential components of a clinical genomics report, including patient information, clinical context, methodological details, variant interpretation, and recommendations. It highlights the importance of clarity, precision, and acknowledgment of technical and interpretative limitations, ensuring that genomic findings are effectively translated into actionable clinical insights.

Keywords Clinical report · Variant prioritization · Clinical relevance · Diagnostic interpretation · Rare diseases

The Objective of a Clinical Genomics Study in the Context of Rare Disease Diagnosis

Let's revisit the objective behind conducting a clinical genomics study in the context of Rare Disease diagnosis. As we have previously mentioned, in this case, the ultimate goal is to find in the patient (or the family) the variant (or variants) that explain, and are

M. A. Marti, *Clinical Genomics*,
https://doi.org/10.1007/978-3-032-34667-4_8

therefore causal, of the observed pathological phenotype. In other words, what we are looking to find—and then report—is the pathogenic variant(s) that explain the presumptive clinical diagnosis. However, if we focus too narrowly on this specific objective, the vast majority of clinical genomics studies would turn up negative. That's why, seeking a broader perspective, we are going to shift the goal slightly to the following: "to find and report those variants that have"clinical relevance" for the case or patient under study.

Now, let's analyze what characteristics a variant must have in order to be considered clinically relevant. In descending order, it is clear that if the variant "is" diagnostic, it is clinically relevant. A diagnostic variant will be, in this case, a variant classified as pathogenic in a gene where the expected phenotype matches the one observed in the patient. Similarly, if we find a variant classified as likely pathogenic, in the same gene, we can say that we have a variant that is potentially diagnostic and this should also be reported, because it is clinically relevant. These are the clearest cases and this is what is usually considered a positive or diagnostic result.

The subsequent cases, which in our view are also clinically relevant and therefore should be reported, nevertheless have a higher level of uncertainty than the foregoing. The first of these cases is when the best variant we find is in a gene whose expected phenotype matches the one observed, but the variant is classified as a Variant of Uncertain Significance. As mentioned in the previous chapter, this should be a VUS without contradictions, possessing pathogenic labels, but altogether not sufficient to classify it as likely pathogenic. In this case, we suggest reporting the variant and treating it as a "Working Hypothesis" or "Diagnostic Hypothesis." What we have is a variant that, in order to be definitively diagnostic, must be confirmed as pathogenic. Usually, to advance in these cases, it is suggested to advance in the analysis of family segregation of the variant, which may result in its reclassification as (likely) pathogenic, and/or in some cases, to proceed with functional studies.

A second, different case, comprises those patients in whom a (likely) pathogenic variant is found, but in a gene for which the expected phenotype does not completely overlap (or coincide) with the observed one. In this regard, it should be considered that sometimes phenotypes are complex, there can be a wide heterogeneity in them, and they also develop as the patient grows, so it is possible that the overlap is not total. In these cases, where the variant can explain part of the observed picture, it is also recommended to report the variant. Clearly, if the same occurs with a VUS, its report is also possible but should be analyzed with more detail and precision the possibility that it explains, in case of actually being pathogenic, a significant part of the observed phenotype.

Finally, and to complete the picture, we suggest two further cases where we consider that the variant has clinical relevance and should be reported, although these are rare cases in the context of genomic studies for Rare Disease diagnosis. Furthermore, in these cases, the variants do not classify according to ACMG criteria. The first case involves the detection of a variant that may impact therapeutic decision-making, specifically regarding a potential treatment. These cases generally comprise variants of pharmacogenomic relevance, and while it is possible to detect them in an exome/genome, a specific annotation and analysis must have been done with this purpose. The second involves the detection of one or more variant(s) related to potential prevention strategies. These variants are called risk variants (which we will analyze later), and they too must be specifically annotated and analyzed with this purpose.

When to Report and When Not to Report a Variant

Having analyzed in detail which variants initially have clinical relevance, we can then reflect on which variants should be included in a clinical genomics report. It is clear that in a clinical genomics study whose objective is to seek a molecular diagnosis, we must first report those variant(s) that can be considered diagnostic. This certainly includes the first two types of variants mentioned in the previous section, but in a broader view, it would suggest including all those variants that, according to the criteria previously described, have clinical relevance.

Perhaps more difficult than determining which variants "to" report is determining which variants "not to" report. It is important to understand which variants not to report, given that, for example, in an exome, and not to mention a genome, the number of variants is enormous. In this case, it is important to emphasize that variants that do NOT have clear clinical relevance should not be reported. It is not usually useful, and it can even be counterproductive, for the health professional to receive a report with one (or more) variant(s) whose clinical relevance is doubtful or even nonexistent. This includes, on the one hand, (likely) pathogenic variants that are found in genes that do not have a link with the observed phenotype, and on the other hand, variants in relevant genes, but which are VUS (or even benign) without a clear possibility of becoming pathogenic through a further study of the case and/or the variant and the consequent inclusion of more pathological labels. This philosophy of not reporting variants with questionable clinical significance necessarily results in the production of negative reports. In this context, we refer to a negative report as one for which no variant with clinical relevance has been found, and therefore, none is reported.

It is important to note at this point, according to statistics provided by the analysis of multiple clinical genomics projects, that currently (2025) the percentage of negative reports is slightly more than 50%, and the percentage of cases with a molecular diagnosis with high confidence, that is the detection and reporting of a (likely) pathogenic variant, is close to 30%. Of course, these percentages vary significantly for different groups of pathologies, being higher for pathologies with less heterogeneity of loci and more history, such as hemoglobinopathies, growth disorders, or metabolic disorders; and slightly lower for more heterogeneous pathologies, both in relation to the symptoms observed and the causal genes, such as pathologies associated with intellectual development or immunological disorders.

Roadmap for Variant Prioritization in Rare Disease Diagnosis

Having established which variants have clinical relevance, which are those that we must report, within the framework of the classification and tags established by the ACMG, we will now propose and describe a strategy for prioritizing variants that allows for an efficient analysis of the case.

The first step consists of, starting with the medical history and/or presumptive clinical diagnosis, organizing the genes into groups. Typically, there will be a primary group of genes, which we will call "candidate genes" or "main panel," which are those whose expected phenotype due to the presence of pathogenic variants in them, clearly matches the phenotype observed in the patient. This list of genes can be provided by the professional and/or can be compiled from the analysis of commercial panels for the pathology and/or from literature analysis. Nowadays, there are also various computer tools (such as GenPhenIA, Phenomizer or FindZebra) that, starting from the symptoms, will deliver a list of candidate genes. This main panel usually consists of between five and twenty genes. For some pathologies that present extensive heterogeneity of loci and symptoms, a second list of genes can be compiled, which we can call a secondary or extended panel. These genes will be those that are related to the observed symptomatology and/or the presumptive clinical diagnosis but to a lesser extent than those in the main panel. For example, those where the overlap between the observed and expected phenotype is partial. Typically, the extended panel will comprise between 50 and 100 genes along with the main panel, although for some pathologies it may extend to 500. Finally, the rest of the genes will be held in a third group.

Once the genes are organized, we proceed to the second step. This consists of first selecting all those variants that: (i) are found in the main panel and (ii) are classified as (likely) pathogenic. If the case presents one (or more) variants that meet this criterion, we will then review them in detail. Specifically, we must check three things: (1) that the variant is correctly classified. This includes reviewing and corroborating that the labels placed are correct. (2) That the expected phenotype coincides with the one observed. Although this step seems redundant since the variant is in the list of candidate genes, it is possible that the list was obtained from a commercial panel and/or from literature work without having done an analysis related to the symptoms of the particular case. It is for this reason that at this point the relationship of the expected symptoms with those specifically reported for the case should be reviewed. On the other hand, it never hurts to double-check. (3) Review that the variant has indeed been called correctly. To do this, we must review the BAM file (where reads are mapped to the reference genome and variants were called, see Chap. 3) and see that the variant can be observed in the reads and that the number of reads in which the variant is observed (its read depth) is consistent with the zygosity of the variant. If the variant "passes" these three analyses, then it will be reported and we will have achieved a (potential) molecular diagnosis.

In the case that there are more than one variant that meets the above mentioned characteristics and in different genes, the labeling must be reviewed since its labeling is conditioned by the presence of another variant(s) that explain the case. Usually, there will be a gene that has one (or two variants in the case of a compound heterozygote) that better explains the clinical picture, with which the second variant is likely to decrease its pathogenic potential and perhaps be reclassified as a VUS. In this case, the second variant should not be reported. In the case that the variants of both genes can explain the observed phenotype, or part of it, we can consider that both have clinical relevance and report them all. It is important at this point to reinforce that, if there is, a (likely) pathogenic variant that explains the case, it is very unlikely that a VUS type variant in another gene is clinically relevant, therefore in the vast majority of cases, there will be no need to report them.

In the event that no variant meets any of the earlier specified characteristics two alternatives arise. In the first case, it's necessary to search for (likely) pathogenic variants in the secondary panel, and if one is found, proceed to analyze it thoroughly to confirm it and determine whether it has, or does not have, clinical relevance and therefore whether it should be reported (or not). Concurrently, what must be done is to analyze the presence of VUS variants in the main panel (or possibly the secondary panel). It is almost certain that variants with these characteristics will appear, and it must be assessed whether they comprise a diagnostic hypothesis and therefore should be reported, or whether they do not possess clinical relevance, in which case the case result could likely be negative.

Initially, we must focus on those VUS that: (a) do not present contradictions and are close (in terms of the number of labels) to being possibly pathogenic, or (b) present contradictions that can be resolved. For both variants, we will analyze the variant in depth according to the three points already mentioned: (1) correct labeling and classification, (2) concurrence with the patient's phenotype, and (3) correct calling of the same. If these requirements are met, the variant is to be reported.

In this context, the situation may arise where a VUS is found in the main panel, and a (likely) pathogenic variant in the secondary panel. In this case, a comparative evaluation of them must be carried out in terms of: (i) their capacity to explain the observed phenotype, concordance between the observed and expected phenotype, and (ii) the level of evidence of their pathogenicity, the quantity and quality of the labels that determine their ACMG classification. Only if both variants present equivalent evidence should both be reported, whereas if one is clearly better, or in other words has significantly more pathogenic potential, only that one should be reported.

Finally, if no (likely) pathogenic variants are found and there is no VUS that constitutes a good diagnostic hypothesis, a negative report must be issued. In this context, two situations worthy of attention arise. The first involves determining how many VUS should be analyzed as possible diagnostic hypotheses before deciding that the case will definitely be a negative result. While the total number of VUS is usually small when the candidate gene panel is small, in some pathologies with expanded panels, the VUS can number in the hundreds. In these cases, to perform an efficient job and not waste time analyzing meaningless variants, it is advisable to have a predefined strategy based on the analysis of a fixed number of VUS-type variants (usually between 3 and 5). Once the number is defined, the VUS are ordered according to their level of evidence, how many and what type of labels they possess, such that those ranked highest have the most labels of a pathogenic nature, or more strongly pathogenic labels (thus PS or PM over PP). Once all the VUS are ranked in this manner, the first will be analyzed in relation to the three aforementioned points. If, after this analysis, the variant is considered to have clinical relevance, then the second is analyzed and if this second one does not, the first is reported and the case is closed. If the first variant is not clinically relevant, the second is analyzed in the same way, and if this is also not clinically relevant, proceed to the next one. In this way, the analysis continues until the predefined number of unreportable VUS have been analyzed, and upon reaching this number, a negative report is issued. The logic behind this technique, which works very well for only two or three variants, is that since they are ordered according to their level of evidence, the likelihood that as we move down the ranking, their clinical relevance decreases significantly; therefore, if the first few are not reportable, the probability that the subsequent ones will be approaches zero very quickly.

The second significant situation when issuing a negative report involves trying to verify that none of the variants with the greatest explanatory power for the patient's phenotype are actually present in the sample. In other words, we want to verify that at the positions where the known pathogenic variants, which could explain the condition, are located, the patient is indeed homozygous for the reference genome. Why must we check this? Because, it is always possible that the patient has a variant that obviously explains the case, a pathogenic variant in one of the candidate genes, but that it was not called, either because there was a problem in the variant calling and/or in the sequencing process that resulted in low depth at the corresponding position. How can we verify this possible mistake? First, given the candidate genes, we must have a listing of the (likely) pathogenic variants known in the gene. These variants are, according to what we've read, those that have the highest chances of explaining the condition. Then, once this list is available, it should be verified that the patient is homozygous for the human reference genome at each of these positions and that they exceed a minimum depth threshold (for example, 10×). If this is the case, we can be reasonably sure that the sample does not possess one of these variants and that the genomic analysis (plus the report) is clearly negative. If a position is found, for example, that was not read properly where there is a known pathogenic variant, it is important to report this in the negative report and possibly perform a complementary Sanger sequencing to definitively rule out the absence of this variant.

Confection and Elements of the Report

Having thoroughly determined which variants we must report when detected, and what to do in cases where none was found, and thus issuing a negative report, we will next examine all the essential components of a clinical genomics report (Fig. 8.1) and the reasons for their inclusion.

Firstly, the report must include basic identification data, such as the institution and the professional overseeing the case, along with patient identification. If the case is part of a research project and it is anonymized—meaning it lacks personal data—the patient identifier will correspond to an identification number, which enables the professional and institution to link the report with the personal data of the patient and their family.

Secondly, the report must contain the Clinical History, Presumptive Clinical Diagnosis (if any), and the reasons underlying the study's request as expressed by the attending professional. This information is extremely relevant, especially if the case requires review months or years after its conclusion, since it facilitates a re-analysis of the case, including, for example, a re-annotation of the variants, in line with the strategy outlined in previous sections.

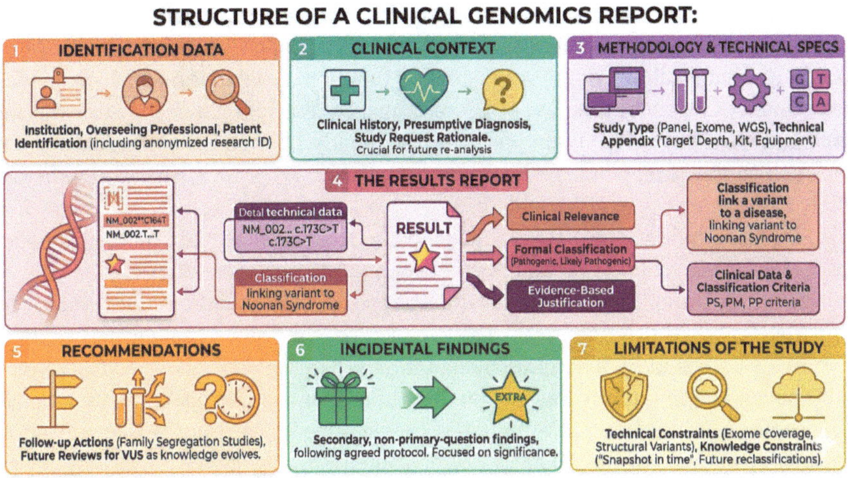

Fig. 8.1 Structure of a clinical genomics report

Thirdly, the report should state the type of genomic study that was conducted, whether a panel, exome, or genome. Additionally, it is suggested to include technical data (target depth, capture kit, equipment used, etc.) in an appendix.

The fourth item, and the most important, is the results report. In the results, we must initially report the variants that have passed the prior analysis, those with clinical relevance. For each variant, the following should be reported: (i) technical data such as chromosome and position, its nature (for example C>G), in which gene it is located, its zygosity, its depth, and the depth at the position. Furthermore, if possible, a link to the read mapping (BAM) file should be provided where the variant can be observed in the mapping. (ii) a comprehensive classification of the variant must be provided, which includes each of the labels it possesses, with a brief description justifying the assignment of each, and finally its classification. Secondly, a biological and clinical description of the variant's impact in the context of the gene should be made, depending on the nature of the variant, the function of the gene, and the knowledge available from other cases. Finally, as a summary, the analyst may provide a conclusion, concisely stating if the variant is considered diagnostic and/or to which of the criteria for assigning clinical relevance it corresponds.

An example of this section could be as follows: The variant NM_002524.5:c.173C>T - p.(Thr58Ile), detected in heterozygosity in the NRAS gene (MIM: 164790), corresponds to the substitution of a C for a T in the coding region (ENST00000369535.5) of exon number 3, resulting in the substitution of the Threonine residue at position 58 with Isoleucine in the resulting protein.

This is a missense variant at a GTP-binding site, located in a mutational hot spot (PM1). It is a missense variant in a gene where missense variation is a mechanism of hyperactivation and disease development (PP2). This variant has a low population frequency in the GnomAD database (PM). It comes with a pathogenic prediction based on 12 pathogenic predictions and no benign predictions (PP3). Reported in ClinVar as pathogenic associated with Noonan Syndrome (OMIM: 613224) and in LOVD as likely pathogenic (PP4). It shows a functional evaluation which concluded that it induces a gain of function by promoting a conformational change towards the constitutively active GTP-bound form, consequently causing the activation of ERK and AKT at the basal level (PS3). Based on these labels, the variant is classified as likely Pathogenic (1PS, 1PM, and 3PP).

Gene Description: The NRAS gene is located on chromosome 1p13.2 and codes for a membrane protein that shuttles between the Golgi apparatus and the plasma membrane. This shuttling is regulated through palmitoylation and depalmitoylation by the ZDHHC9-GOLGA7 complex. The encoded protein, which has intrinsic GTPase activity, is activated by a guanine nucleotide exchange factor and inactivated by a GTPase-activating protein. Variants in this gene have been associated with Noonan Syndrome (AD, MIM: 613224), which is characterized by distinctive facial dysmorphia, short stature, congenital heart defects, and a variable degree of developmental delay. Other findings may include a broad neck, an unusual chest shape with a superior pectus carinatum and an inferior pectus excavatum, cryptorchidism, varied coagulation defects, lymphatic dysplasias, and ocular abnormalities.

Conclusion: The expected sign and symptomatology for a pathogenic variant in NRAS aligns with those described for the patient, which implies that the found variant is classified as likely pathogenic, it is considered likely diagnostic for Noonan syndrome.

Fifth, a section with recommendations can be included, such as the analysis of the variant segregation in the family group, reviewing the patient in relation to an expected but unobserved symptom, or, in the case the reported variant is a VUS, recommending a review after a certain time to analyze the possible emergence of new evidence.

Sixth, and if it was agreed with the treating professional and is within the work protocol (as we will see later), incidental findings should be included. These should follow an agreed, systematized, and predefined format similar to that of the results but with a special focus on their clinical significance and associated recommendations.

Finally, the last part of the report should address the limitations of the study. Like any medical study, a genomic study has limitations that are inherent to the technique, such as methodological limitations due to the knowledge available at the time of the study, and specific limitations associated with the type of study that was specifically performed. For whole exome studies, for example, a limitation is that not really 100% of the exome is covered, thus there remains a possibility that variants causing the patient's condition may not be detected with the assay. For the case of exomes and panels, it should be specified that the study is not optimized for detecting balanced structural rearrangements (inversions, translocations, etc.), or trinucleotide expansions. It should also be indicated that the study may not be conclusive if the variants are present in germline mosaicism, or are restricted to a specific tissue. Regarding the timeliness of the study and the knowledge available on the day the report is made, it is important to mention that it is based on the information available on that date from biological databases and scientific publications. Thus, there might be a variant currently unlinked to the patient's phenotype, which in the future, could be reclassified as causing the disease. Highlighting these types of limitations is crucial, especially when the report results in a negative outcome.

As we conclude this exploration of rare disease management through clinical genomics, it marks a pivotal transition in our narrative. Thus far, we have delved deep into the intricacies of diagnosing and understanding rare genetic disorders, enriching our knowledge and refining our approach towards these complex conditions. Now, we turn towards broader horizons—extending into the realms of population genetics, Ethical, Legal, and Social Issues (ELSI) in genomics, and the nuanced field of cancer genomics. The upcoming chapters promise to broaden our understanding of how genetic diversity shapes population health, explore the ethical quandaries posed by genomic information, and dissect the genetic underpinnings of cancer. This progression underscores the versatility and expansiveness of Clinical Genomics, illustrating its profound impact across various facets of medicine and public health.

References and Suggested Further Readings

Rehm HL. Evolving health care through personal genomics. Nat Rev Genet. 2017;18(4):259–67. https://doi.org/10.1038/nrg.2016.162.

Vairo FP, et al Defining the clinical value of a genomic diagnosis in the era of next-generation sequencing. J Annu Rev Genom Hum Genet. 2016;17:419–40. https://doi.org/10.1146/annurev-genom-083115-022348.

Chapter 9
Population Genomics and Genetic Diversity

Abstract This chapter explores the principles of population genomics and their relevance to understanding genetic diversity and complex diseases. It begins with the concept of allele frequency as a fundamental metric for describing genetic variation within populations and introduces population genetics as the framework for studying how these frequencies change across time and environments. The chapter highlights the inverse relationship between allele frequency and effect size, contrasting rare, high-impact variants with common polymorphisms that contribute modestly to disease risk. It then examines polygenic diseases, emphasizing their multifactorial nature and the combined influence of numerous genetic variants, environmental factors, and lifestyle. Genome-Wide Association Studies (GWAS) are presented as the primary method for identifying statistical associations between variants and phenotypes, with particular focus on the interpretation of odds ratios as measures of relative risk. Building on these associations, the chapter introduces Polygenic Risk Scores (PRS) as predictive models that aggregate small genetic effects to estimate individual disease susceptibility. Finally, it addresses critical challenges related to population structure, ancestry bias, and the underrepresentation of non-European populations, underscoring the need for more inclusive genomic research to ensure equitable and accurate clinical applications.

Keywords Population genomics · Allele frequency · GWAS · Polygenic risk score · Genetic diversity

Allele Frequencies and Population Genetics

In the realm of genomics, as we have already seen, allele frequency is a foundational concept crucial for understanding their potential biological role and the broader landscape of population genetics. At its core, allele frequency relates to the measure of

M. A. Marti, *Clinical Genomics*,
https://doi.org/10.1007/978-3-032-34667-4_9

how common a variant of a given gene—referred to simply as an "allele"—is found within a given population. Usually all members of the species are considered, thus human variant allele frequencies measure the fraction of human genomes harboring this particular variant. Allele frequencies are determined by analyzing the variants that are present in a sample of individuals. Because humans are diploid organisms—each individual possesses two copies of each gene, one inherited from each parent—therefore, the calculation of allele frequencies involves counting alleles rather than individuals. If a variant is present in an individual in heterozygosis it will contribute as one allele, while if the variant is in homozygosis it would contribute with two alleles. Let's consider an example, in which we are assessing the frequency of a variant allele, Chr1:posXX C >T, within a gene sample from a population of 100 individuals. If 70 individuals possess the variant in heterozygous form, 20 are homozygous for the variant, and the remaining 10 do not present the variant, the total variant allele count is $70+2\times20=110$. Dividing this number by the total number of alleles (200 in this case) gives the resulting Allele frequency of 0.55. Population genetics, is the study of these allele frequencies. how they are distributed among different populations, how they change over time and space, and how they are related to many biological phenomena including diseases. Massive Sequencing Technologies have revolutionized this field by the sheer amount of data they provide. As described in previous chapters, after the completion of "The" human genome, several projects started to study human genomic diversity, and thus our population genetics. In relation to disease, possibly the most important observation related to population genetics, is the already mentioned inverse relation observed between allele frequency and effect size. When we focused on Mendelian Diseases, we looked at those variants displaying very low allele frequencies, with a very large (close to infinity) effect size. In other words, these low frequency variants, were almost deterministic or causal in relation to disease. In this chapter, we will look at the other end of the spectrum. We we will study those variants displaying moderate to large frequencies (1%–50%), usually referred to as polymorphisms and how they are related to diseases such as cardiovascular, metabolic, or neurodegenerative among others. These variants, we will show, have very small effect sizes, and thus many variants have to be analyzed to have a clinical significant contribution, in the context of complex polygenic models. The relation between high frequency variants, or polymorphisms and disease, is studied in the so called Genome Wide Association Studies (GWAS), which we will also discuss below.

Polygenic Diseases

Polygenic Diseases starkly contrast with monogenic disorders. Unlike illnesses derived from a single gene mutation, these genetically complex diseases arise from multifaceted origins—interactions among genetic variants, environmental factors like exposure to chemicals or infections, and lifestyle choices such as diet, exercise, and smoking all play vital roles. For instance, the prevalence of type 2 diabetes increases not just with genetic predisposition, but also with factors like high-sugar diets and sedentary lifestyles. Through advancements in genomics, we now understand that these widespread conditions like cardiovascular diseases, obesity, Crohn's disease, Alzheimer's, and even certain cancers are influenced not by one malfunctioning gene, but by a network of genes each contributing subtly and cumulatively. These small genetic variations create a baseline of an individual's risk, which, when combined with clinical data, paints a comprehensive picture of potential health outcomes. Importantly, these are the disorders most prevalent, or common, in the population, standing in clear opposition to rare genetic diseases. In the context of Clinical Genomics physicians could use this knowledge to tailor interventions more accurately to suit each individual's genetic profiles, steering away from the one-size-fits-all paradigm that dominated previous medical practices. This shift not only enhances disease management but also elevates the preventative strategies by allowing early interventions tailored to personal genomic landscapes.

To exemplify the impact of Clinical Genomics in this scenario, consider the case of a patient whose routine blood tests reveal elevated lipoprotein levels. Traditionally, a physician might view this result as minor and recommend standard interventions such as dietary modifications or increased physical activity. However, the integration of genomic analysis into the patient's evaluation can dramatically transform the clinical approach. Let us suppose the patient's genomic data includes a C→T variant in the APOA gene, which is known to quintuple lipoprotein concentrations, and another variant in the LDL receptor gene that triples the risk of coronary disease. Suddenly, the previously not so relevant lab result, appears as a possible indicator of a chronic, genetically-driven metabolic disorder. This revelation could prompt more aggressive interventions, perhaps including specialized medications and more frequent monitoring, tailored to manage the heightened risk effectively.

It is also important to remark, that from a clinical perspective, these analyses are most relevant during adolescence and adulthood. In contrast to most Mendelian diseases which are manifested in early life, polygenic and multifactorial diseases tend to arise later and exert their greatest impact during adult life. Therefore, the optimal window for implementing genetic risk assessment strategies is in young or middle adulthood, when genetic predispositions begin to interact more clearly with environmental and lifestyle factors. Moreover, since there are many different polygenic diseases, determining which is the one, or the few for which each individual is at higher risks allows focusing the preventive efforts, in time and money towards them.

This shift to a genetics-informed perspective raises several crucial questions: How do we ascertain these risk multipliers? Where do these associations between genotypes and phenotypes stem from, and just how predictive and reliable are these findings? The answers to these questions are typically derived from Genome-Wide Association Studies and the development of the Polygenic Risk Scores.

Genome Wide Association Studies (GWAS)

The associations between genetic variants and pathologies stem from what is known as genome-wide association studies. These studies aim to first identify statistical relationships linking genetic variants, usually those with relatively high allelic frequencies

(1%–5%) to a given phenotype, which in our case of interest is a disease, or some key symptom of it. Following these associations, the analysis delves deeper into understanding the underlying biological phenomenon that explains the observed genotype (variant)-phenotype relationship, which we already know is one of the key themes in clinical genomics.

Essentially, the most common design of a GWAS consists of comparing two groups of individuals: a case group (people affected by the disease under study) and a control group (healthy people). The genomes of both groups are sequenced—or at least a large portion thereof—and millions of genetic variants of each individual are identified. Subsequently, the allele frequencies of these variants are compared between the case and the control group to look for significant associations, for example, noting that "the majority of cases carry a given variant" while "in the control group, the majority of the individuals are homozygous for the reference genome, that is, they do not possess the variant."

This comparison is conducted via a contingency table, in which the frequencies of each allele in both groups are recorded. From this table, various statistical tests (such as the False Discovery Rate) are performed to quantify the statistical significance of the association (its p-value). For each genetic marker—that is, for each variant probed across the genome—it is calculated whether or not the association is statistically significant. This is displayed in a graph known as a Manhattan plot (so named because it resembles a typical photo of the Manhattan skyline), a chart where all the genome markers are shown on the X-axis and the level of statistical significance (*p*-value), expressed in units of negative logarithm, on the Y-axis. Generally, the significance threshold is set around $p = 1 \times 10^{-7}$, since about one million hypotheses are being tested simultaneously. This value is approximately equivalent to a p-value of 0.01 in a traditional study, but adjusted for multiple comparisons. In summary, in a GWAS, for an association to be considered truly significant, the p-values must be lower than 1×10^{-7}. Everything else you will see in the literature—improvements, statistical refinements, Bayesian methods, etc.—are variations on this basic scheme.

Once a statistically significant association is identified, the most delicate part follows: interpreting its magnitude. Most genome-wide association studies—not to say all—do not merely assert the existence of a statistically significant association, but also calculate the previously mentioned effect size, or more specifically, the odds ratio (OR).

The Odds Ratio

It is important to understand what the OR actually measures. The odds ratio compares the chance (or "odds") of belonging to the case group—for example, sick people—if one possesses a certain allele (for example, allele G), in comparison to the chance of belonging to the same group if one possesses the alternative allele (for example, allele A). The allele with a larger than 1 OR is called the risk-allele, and usually must be different from the reference allele. In other words, the OR estimates how many times more likely it is to be among the cases if one carries the variant, compared to those who do not. Therefore, it is very important to note that the OR is always referred to a reference allele and a reference population. The value obtained—whether it is 1.5, 3, or 0.8—does not have an absolute meaning, but it expresses a relative likelihood, relative to the reference population and allele.

To better understand what "chance" means in this context and how it differs from a real "probability," let's think about a simple example. If we roll a fair die and ask what the probability of rolling a one is, we all know that it's 1/6. This probability is deduced from theory: the die has six faces, all with an equal chance of appearing. In this case, probability and chance coincide because the system is ideal and the conditions are known. However, if the die were loaded or biased, we could no longer calculate its probability theoretically. What we could do is measure empirically how often a one appears after many rolls. If, for instance, a one comes up 30 out of 100 times, we would say that the observed chance is 0.3, which is an empirical estimation of the probability. Similarly, in GWAS, the odds ratio functions as this type of empirical estimator: it is calculated from the data observed in a contingency table, not from a theoretical model. That is why we say that the OR is an empirical estimator of the relative risk, based on real observations within a population sample.

Now, let's look at another example to understand why the odds ratio is so important. Imagine you go to a store and buy a lottery ticket. I ask you:—"What is the probability that you'll win the lottery?" Well, probably you can't provide an answer, since you don't know how many tickets were sold. In theory, there could be one million possible numbers, but if you don't know how many were actually sold, or how the draw is defined, you can't calculate the real probability of winning. Now suppose you buy one number, and your neighbor buys two. You still don't know the exact probability of winning—because the total number of tickets sold is still unknown—but you do know who, you or your neighbor has more chances: He has! And you know how many more times he is likely to win, compared to you: twice. This comparison is possible without knowing the absolute probability, just by observing the relationship between each person's "chances."

The same applies to the odds ratio. We cannot know the exact probability of developing the disease being a carrier of the risk allele (in our example, G), nor the exact probability if one has the reference A allele. But we can estimate how many times greater the chance of having the disease is in one group compared to the other. For example, if the OR is 6, that means that individuals with the G allele are six times more likely to develop the disease than those with the A allele. In other words, the odds ratio serves as a relative comparison of chances, similar to comparing the number of bought lottery tickets without knowing the exact probability of winning. It does not tell us if someone will "win"—that is, become ill—, but it does indicate who has more tickets in the draw, and thus who has a higher chance of developing the disease.

An interesting additional issue that needs to be mentioned, is that the odds ratio is defined for a single allele. However, we already know that each of us carries two alleles per locus, one inherited from the father and one from the mother. Therefore, the question is: if a heterozygous person (with just one risk allele, for example "G") has a certain risk, what happens with someone who is homozygous and has two G alleles? Is his/her risk doubled, or how do the two alleles combine their OR? The answer depends on the genetic model adopted, since complex diseases do not have a clear Mendelian pattern of dominance or recessivity. Nonetheless, approximate models are usually used to estimate how the effects of both alleles combine. There are mainly two models. The Multiplicative model, where the risk associated with having two alleles is the square of the risk associated with having just one. This is the most commonly used model, because it accurately reflects the behavior observed in most studies. And the additive model, where the risk with two alleles is simply twice the risk with one. Most GWAS use the multiplicative model, which tends to increase the estimated risk when combining the effects of the two alleles. Nevertheless, odds ratios usually display values close to 1, so the differences between the two models are rarely dramatic.

Finally, it is worth mentioning that not all GWAS are limited to comparing cases and controls. There are also quantitative GWAS, where an odds ratio is not calculated, but a beta value (β), which indicates how much a quantitative characteristic (for example, height or cholesterol level) increases or decreases for each additional risk allele. Unlike the OR, the model with beta is usually additive, although the logic of interpretation is very similar: in both cases, the goal is to estimate the effect of alleles on a trait or disease.

The information derived from all the performed GWAS is made publicly available in a free and public database called the GWAS catalog. There, researchers can search by variant, gene, or phenotypic trait, and find for each case the allelic frequency, the odds ratio or beta, the sample size, and the statistical significance of the finding. The nomenclature used in the catalog may seem complex—it relies on identifiers such as rsID (for example, rs429358 for ApoE4)—, but it follows a standard that allows the integration of data from thousands of studies and their comparison. In the GWAS Catalog, each variant is identified by a code that begins with "rs," such as rs429358. This "rs" comes from the term reference SNP and corresponds to a unique genomic position identifier assigned by the dbSNP database. Therefore, every reported variant must be listed with this prefix—"rs" followed by a number. It may seem like an unfriendly nomenclature, but it is the historical and universally adopted standard for referring to single nucleotide polymorphisms.

Each catalog entry displays several key parameters: The statistical significance value (p-value), which in some cases can be extremely low, for example, 2×10^{-28}, indicating a highly robust association. The frequency of the risk allele (Risk Allele Frequency, RAF), which can be high or low depending on how common the variant is in the analyzed population. The odds ratio (OR) or, in the case of quantitative traits, the beta value (β). For example, an OR of 9.6 represents one of the strongest known associations—a nearly tenfold increase in the chance of presenting the trait. In quantitative studies, a beta value of 1.06 could indicate an average increase of 1.06 units in the measured characteristic (for example, milligrams per milliliter of vitamin D), although the exact units should be consulted in the original article.

Additionally, the catalog provides information about the gene closest to the variant—which may be within a gene or in a nearby regulatory region—along with the associated characteristic or phenotype and the details of the original study in which the association was reported. From there, the researcher can delve deeper into the corresponding study to analyze the variants, their biological context, and their functional relevance in more detail.

Polygenic Risk Scores (PRS)

Once the variants associated with a trait or disease are identified, the next step is to construct a predictive model of the total genomic risk, known as Polygenic Risk Score (PRS). The goal of the PRS is to estimate individual genetic risk from the combination of many variants, each with a small effect, distributed throughout the genome. The steps to generate a PRS are the selection of variants, construction and calibration of the model, validation of the model and finally, its application in the clinical context. Let's analyze each of them briefly. The first step involves the selection of the variants to be included in the model. These are chosen based on the results of the GWAS, using criteria such as the statistical significance of the association (*p*-value), the frequency of the risk allele, and the effect size (odds ratio or beta). The reference population is crucial, as we will see later; it is essential that the GWAS used correspond to a genetic ancestry similar to that of the population where the model will be applied. It is also necessary to avoid double-counting by correcting for linkage disequilibrium. In simple terms, if two variants are highly correlated, usually because they are within the same gene or region, only one of them should be chosen as the representative marker for the PRS model. Once the variants are selected, the model, which is a mathematical formula to combine their effects is defined. In simpler models, the PRS is calculated by combining the estimated effects of each risk allele present in the individual, weighted by their odds ratio. Most models also have additional parameters that allow some adjustment or training. The model must be trained on a large population where both the genotype and phenotype of individuals are known. The sample sizes required are usually large—on the order of thousands or hundreds of thousands of individuals—, which is why most current models have been developed using large-scale biobanks, such as the UK Biobank or the All-of-us project, which gathers genomic and clinical information from more than half a million people. Once trained, the model must be validated on an independent population to verify that the predictions match the observed data. It is important to understand that polygenic risk score is a probabilistic outcome, not a nearly deterministic outcome usually observed for Mendelian disorders. For example, a person in the 99th percentile of the PRS for coronary artery disease may have twice the risk of the average population, but not a certainty of getting sick. There is always an overlap between healthy individuals and those affected. In other words, the PRS indicates greater predisposition, not destiny.

Finally, it is crucial to recognize how individual and population interpretations of a PRS result are to be approached in clinical genomics. The first notable point is that the obtained PRS should be understood in two complementary fashions. At a population level, PRS enables the stratification of individuals into distinct risk groups. Identification of such a group facilitates targeted monitoring and intervention, enhancing the efficacy of preventive health measures. At an individual level, this data helps to place a person within a specific risk percentile relative to the general population for each analyzed disease. Therefore, for those diseases where its genome results in an increased risk healthcare providers might advocate for specific preventive actions, which will therefore be tailored to the individual's risk profile. As an illustrative example, if an individual has an average risk of cardiovascular diseases, but shows a higher risk score (PRS) for type 2 diabetes, the preventive strategy would prioritize metabolic and lifestyle interventions aimed at improving insulin sensitivity and glucose regulation such as reducing refined carbohydrate intake, increasing aerobic and resistance exercise, and performing periodic glucose tolerance tests. Conversely, an individual with a high cardiovascular PRS but low genetic risk for T2D might benefit more from interventions targeting lipid metabolism and vascular health, such as cholesterol-lowering therapy (statins), blood pressure control, and dietary measures focused on reducing saturated. These contrasting examples illustrate how polygenic risk profiling can refine preventive medicine by tailoring surveillance and lifestyle recommendations to each person's specific genetic vulnerability.

Integration of Genetic and Clinical Risk in Medical Practice

One of the most promising applications of polygenic risk scores, and the future of clinical genomics, is their integration with traditional clinical factors to optimize personalized prevention. Various authors suggest that studying the genetic contribution to different diseases allows focusing on preventative efforts, whether through lifestyle changes (active prevention), or through more frequent and targeted medical follow-up. This is a general way to incorporate polygenic risks into clinical practice, but there are also more specific approaches focused on particular diseases. As an illustrative example, lets analyze cardiovascular pathologies.

Imagine a clinical model for coronary disease. In this case, the total risk is calculated by combining measurable well knwon clinical factors—systolic blood pressure, smoking habits, and the different types of cholesterol levels—with the genetic risk estimated by the PRS. Each of these parameters contributes a relative probability of developing coronary artery disease, a relative risk, compared to the general population, which can be summarized as an integrated Odds Ratio. If a person has high cholesterol levels, this could result in an approximately double risk of coronary disease, with respect to the general population, thus an OR = 2. If the individual is also a smoker, the risk is three times the average risk (OR = 3). If their PRS indicates a high genetic risk (OR = 2), the combined risk would be approximately 6 times that of the general population, clearly a very high risk. The models are usually considered multiplicative, assuming that each factor contributes independently. However, in practice, this is rarely entirely true: many factors overlap or are partially correlated (for example, a genetic predisposition can influence cholesterol or blood pressure levels). Therefore, empirical calibration or some type of correction of the model in large local cohorts are essential to avoid overestimating the total risk. The analysis is performed for all patients, and each is then assigned a risk group. Usually three groups, low, average and high risk, are built. Patients at low risk may maintain routine checks or non-invasive preventive strategies. Those at medium risk should uphold standard preventive measures. Patients at high or very high risk require more intensive or invasive interventions, such as frequent cardiological studies or specific pharmacological treatments.

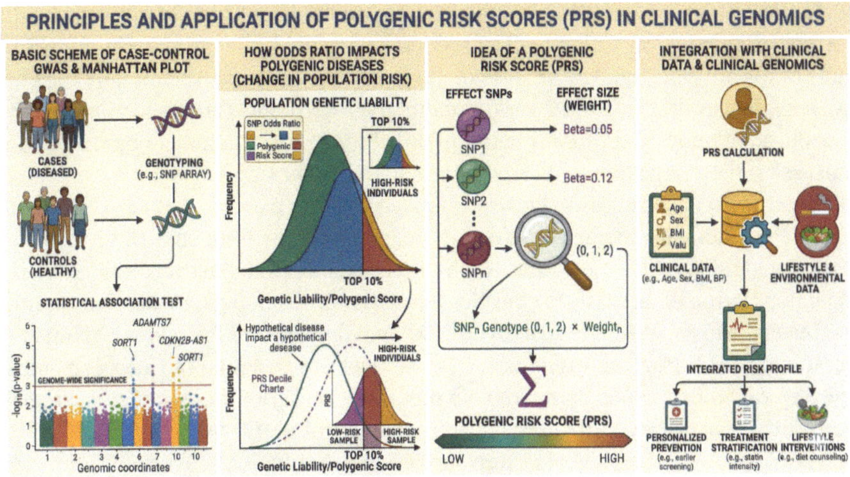

Fig. 9.1 Principles and application of polygenic risk scores (PRS) in clinical genomics. Panel 1 (Discovery): GWAS Panel 2 (Population Distribution): the bell curve of genetic liability. Panel 3 (Calculation): mathematical derivation of a PRS. Panel 4 (Clinical Integration): genetic PRS data is combined with clinical and lifestyle factors to guide personalized medicine

This integration of the genetic risk with the clinical risk allows a clear move towards a personalized medicine, where preventive decisions are tailored to the genomic and clinical profile of each individual. The current challenge is to calibrate and validate these models in diverse populations, ensuring that genetic information genuinely improves prediction and medical decision-making. In the following section we will present some of these considerations as well as some caveats related to GWAS and PRS. Figure 9.1 summarizes the whole process.

A Note on Genetic Diversity and Underrepresented Populations

There are two very important caveats related to the PRS as derived from GWAS, both of which are related to population genetics. The first, is related to population structure. Population structure is one of the main sources of bias in genome-wide association studies. When subpopulations differ significantly in both genetic ancestry and environment, associations can reflect population differences more than real biological effects. As an extreme and illustrative example is the so-called "GWAS of the ability to eat with chopsticks." If such a study were conducted in the United States, comparing random populations, it would likely identify genetic variants "associated" with this ability. However, what the study would actually be reflecting is the Asian ancestry and cultural environment of the participants, not a real genetic predisposition. This type of environmental artifact is common when analyzing populations with different ethnicities, lifestyles, or socioeconomic environments. Thus, studies must statistically correct for population structure. The second key issue, is that as already discussed, odds ratios are calculated in relation to a reference population, specifically the control sample used in the GWAS. Furthermore, the PRS models are also calibrated in a specific population. Therefore, when applying these results to another population, one must ask if their target population is representative of the original study's population. Differences in genetic ancestry between populations—for example, between Europeans and Latin Americans, Asians or Africans—can modify the frequency of variants and alter the validity of the OR. That is why it is essential to adjust risk models to the local population or, at least, to consider that possible discrepancy when interpreting the results. Most GWAS studies (over 75%) and developed PRS were developed in populations of European origin, or having an overrepresenation of European ancestry. Several works, which compared the predictive power of these models with their performance in other populations, show a notable decline. On average, the predictive power of a European polygenic score applied to individuals of African origin is less than 10% of the original, virtually nil. In contrast, Latino American or Amerindian populations retain a higher degree of similarity, yet still exhibit a significant loss of accuracy. This is explained by the fact that genetic diversity in Africa is the highest on the planet—since humanity originated there—, resulting in a much more complex population structure. In Latin America, on the other hand, the population is a genetic mosaic: most individuals have a variable combination of European, Amerindian, and, in some countries like Brazil or Colombia, also African ancestry, reflecting colonial history, migrations, and mixtures that occurred since the 17th and 18th centuries.

In the last years a strong effort related to underrepresented ethnicities has been performed in relation to population genomic studies. However, and although there has been an increase in the participation of non-European cohorts in recent years, the disparity remains significant and does not seem likely to decrease in the short term. This is primarily because most funding and research centers driving these studies are concentrated in Europe and the United States, where the majority populations are European or of European descent. Even in the U.S., where there is an awareness of ethnic diversity, studies involving African American, Asian, or Latino populations remain in the minority. In the future, the most robust solution would be the formation of regional Latin American, and/or African consortia that generate broad genomic databases, with clinical and ancestry information representative of each country. This would allow for the adjustment of predictive models and enhance their local validity. However, this goal remains distant for now, largely due to resource and coordination limitations.

To summarize, beyond the specifics of GWAS, PRS and their clinical and genomics integration, navigating the complexities of these clinical genomics fields reflects broader challenges and imperatives in the world of personalized medicine. It is crucial to recognize and address the nuances of population genetics, environmental factors, and cultural dynamics when applying this state-of-the-art genetic research. The potential to significantly improve health outcomes through tailored interventions based on each individual genome is immense, but it is equally important to ensure that such genetic insights are applicable and equitable across diverse populations. As the field of genomics continues to evolve, the focus must remain on enhancing the predictive accuracy, improving accessibility in diverse populations, and integrating ethical considerations into every step of research and clinical application (more on this in the next chapter). These efforts are vital for realizing the true promise of personalized medicine in improving global health outcomes.

References and Suggested Further Readings

Visscher PM, et al 10 years of GWAS discovery: biology, function, and translation. Am J Hum Genet. 2017;101(1):5–22. https://doi.org/10.1016/j.ajhg.2017.06.005.

Khera AV, et al Genome-wide polygenic scores for common diseases identify individuals with risk equivalent to monogenic mutations. Nat Genet. 2018;50(9):1217–21. https://doi.org/10.1038/s41588-018-0183-z.

Ashley EA, et al Clinical assessment incorporating a personal genome. Lancet. 2010;375(9725):1525–35. https://doi.org/10.1016/S0140-6736(10)60452-7.

Martin AR, et al Clinical use of current polygenic risk scores may exacerbate health disparities. Nat Genet. 2019;51(4):584–91. https://doi.org/10.1038/s41588-019-0379-x.

Chapter 10
Ethical, Legal and Social Issues (ELSI)

Abstract This chapter examines the Ethical, Legal, and Social Issues (ELSI) that have emerged alongside the rapid development of genomics, emphasizing their foundational role in guiding responsible scientific and clinical practice. Originating within the Human Genome Project, ELSI established an unprecedented integration of ethical oversight into large-scale scientific research. The chapter explores key domains, including privacy and data confidentiality, informed consent, equity of access, genetic counseling, and the commercialization of genetic information. Particular attention is given to the evolving models of informed consent, from traditional approaches to dynamic and participatory frameworks better suited to the complexity of genomic data. The discussion of incidental findings highlights the ethical dilemmas surrounding the return of unexpected yet clinically actionable results, balancing patient autonomy with medical responsibility. Legal considerations are illustrated through landmark cases such as the Myriad Genetics ruling, which clarified the non-patentability of naturally occurring human genes. Finally, the chapter addresses the growing challenges of genomic privacy in an era of large-scale data sharing, demonstrating the increasing difficulty of maintaining anonymity. Together, these issues underscore the necessity of integrating ethical reasoning, legal frameworks, and social awareness into the advancement of clinical genomics.

Keywords ELSI · Informed consent · Genomic privacy · Incidental findings · Gene patents

ELSI and the Human Genome Project

Although we already covered the Human Genome Project in a previous chapter, we left out one of its most relevant contributions that go beyond the scientific endeavor and the developed technologies, these are the Ethical, Legal and Social Issues (ELSIs).

M. A. Marti, *Clinical Genomics*,
https://doi.org/10.1007/978-3-032-34667-4_10

In 1988, before the HGP officially began, during these discussions held to define the HGP goals, several participants voiced concern that the power of genetic knowledge could lead to discrimination, stigmatization, and misuse if not handled carefully. In response, Dr. James Watson, then director of the project, proposed that a portion of the HGP's budget (3%–5%) be devoted to studying ELSI. This decision was unprecedented, since no major scientific project had ever embedded ethics within its structure from its inception.

Watson appointed Nancy Wexler as the first chair of the NIH–DOE Joint Working Group on ELSI. Dr. Wexler was born in 1945, the daughter of Leonore Wexler, who developed Huntington's disease in midlife, leading Nancy to live with the uncertainty of being at risk of developing such a devastating disease. This personal connection profoundly shaped her scientific career, and she devoted her life to studying the disease and finding its genetic cause. In the 1980s, she became the driving force behind an international effort to locate the gene responsible for Huntington's disease at Columbia University. She performed leading field research in Venezuela, near Lake Maracaibo, where many extended families had an unusually high incidence of Huntington's. Wexler and her team collected more than 4000 blood samples and family histories, building detailed genealogical maps spanning many generations, leading them to the landmark discovery that the Huntington's gene was located on chromosome 4. Due to the personally experienced emotional, psychological, and social implications of genetic testing, Wexler was unusually sensitive to the ethical dimensions of genetic knowledge and Watson recognized that this experience made her uniquely qualified to guide the HGP ethical and social oversight. Her leadership helped define what ELSI would stand for in the future of Human Genome Research.

The main areas of concern of the HGP in relation to ELSI included: (i) Privacy and Confidentiality: Genetic data unlock deeply personal and predictive information—not only about an individual but also about their relatives. This raises significant questions: Who owns or controls genetic data? Can insurers or employers use this information to discriminate? What measures can be taken to ensure secure storage and restricted access to genome databases? In response to these concerns, the United States enacted the Genetic Information Nondiscrimination Act (GINA) in 2008. GINA aims to prevent the misuse of genetic information in employment and health insurance contexts. (ii) Informed Consent: This facet explores the complexities surrounding how individuals should consent to possibly unknown future research. Solutions such as staged informed consent or the obligation to recontact the donor were developed to address these issues. (iii) Equity and Access: There has been significant debate over whether unequal access to genetic testing and therapies could exacerbate health disparities. Unfortunately, as mentioned in the previous chapter, this remains an unresolved issue; both the scientific community and society at large still face substantial challenges in this area. (iv) Genetic Testing and Counseling: Concerns about predictive testing for conditions like Huntington's disease or BRCA1 mutations have raised profound emotional, ethical, and familial dilemmas. These issues have spurred discussions and developments aimed at ensuring that individuals receive adequate counseling to interpret results. Recommendations on how to manage incidental findings have also been formulated and will be discussed in the next section. (v) Intellectual Property and

Commercialization: Debates continue over whether genes or genetic sequences should be patented. On one hand, patenting can incentivize innovation; on the other, it might restrict access to diagnostic tests. This topic, critical to the balance between innovation and accessibility, will also be explored in further detail in the subsequent sections.

Informed Consent

Confidentiality and Informed Consent At the heart of genomic studies, both in clinical practice and in research projects, lies the informed consent. In the United States, the protection of patient rights and personal health information is primarily governed by the Health Insurance Portability and Accountability Act (HIPAA) of 1996, along with complementary state-level regulations. These legal frameworks aim to safeguard the autonomy, privacy, and confidentiality of individuals in their interactions with healthcare professionals and institutions. HIPAA establishes strict standards for how medical data can be collected, stored, and shared, ensuring that patients maintain control over who can access their health information. Beyond privacy, these laws also reinforce the principle of informed consent, emphasizing that medical decisions must respect each patient's autonomy of will and right to make choices about their own treatment. Together, these protections reflect a broader ethical commitment within U.S. healthcare to balance technological and scientific advances with the preservation of individual dignity and trust. This ethos of conscientious care and consideration extends into the complex realm of clinical genomics. As we navigate through the intricate interplay of genetics and ethics, healthcare systems around the world grapple with issues that aren't just medical or scientific, but deeply personal.

According to the law, for example, for the patient to be able to act according to his/her will, the doctor must "provide information in a clear, sufficient, and adequate manner to the patient's capacity for understanding, about their health status, the studies and treatments that might be necessary, and the foreseeable evolution, risks, complications, or sequelae of the same." In the case of genomic analyses, if we stretch this argument too far, it would imply that the patient understands the risks and implications of all possible findings, from those related to Mendelian diseases to those involving relative risks in complex diseases. Also, the implications of potential findings regarding the risk they pose to direct relatives should be informed and explained. Since even the most basic genetics is not part of the common citizen's knowledge repertoire, achieving the level of understanding required for a truly autonomous decision is currently an insurmountable challenge. If we add the additional cost of transmitting this knowledge from professionals to patients, the problem becomes even bigger. However, the same can be said about many other modern aspects of translational medicine, such as personalized cancer treatment and/or gene therapy; therefore, more realistic models in terms of implementation have been proposed and developed.

The key issues included in the newer or modern concerns, are related to the following key items. (i) Broader consent, where participants give permission for future unspecified research uses of their data and biological samples, under certain governance and oversight safeguards. This type is commonly used in biobanks and national genome projects (like the UK Biobank or All of Us in the U.S.). (ii) Tiered or Layered consent. Where each participant/patient can choose different levels of participation, and return of information (including types of incidental findings). (iii) Similarly, the Dyanamic consent, which comprises an ongoing consent model and continuous dialogue between researchers and participants. Participants can be recontacted to approve or decline new research uses, and they can access updates on how their data are used, as well as decide on the return of results. Together, these modern consent approaches reflect a shift from one-time, transactional agreements toward a more flexible, participatory, and ethically responsive model of engagement. They recognize that genomic research is an evolving process, requiring ongoing communication, transparency, and respect for individual autonomy in the use and governance of genetic information.

Incidental Findings and the ACMG "Black" List

Incidental findings refer to undiagnosed medical conditions discovered unintentionally during studies aimed at other purposes. For instance, a chest X-ray taken for a postural study might incidentally reveal a spot on the lung, indicating a potential pulmonary tumor. While such findings could seem like strokes of luck facilitating early detection of potential health issues, they became, and still are, one of the fears of Clinical Genomics. As shown throughout the book, our genome holds a significant potential for providing incidental findings, with a high variance concerning the certainty and severity of the discovered conditions.

For example, suppose you have a clinical diagnosis for a Mendelian disease, and your doctor suggests sequencing your exome or genome to identify the responsible gene and variant. Upon receiving results, despite finding (or not) the pathogenic variant responsible for the clinical diagnosis, it turns out you carry a mutated BRCA allele with a high risk of cancer, or you are homozygous for ApoE4, carrying implications for developing Alzheimer's disease. Should the doctor inform you of these results? Would you want to know?. Or imagine you decide to sequence your genome with your doctor because of your family history, to analyze the risks for various cardiac and metabolic diseases like diabetes, but it turns out you have an allele responsible for a severe Mendelian disease manifesting in adulthood, with a dominant inheritance pattern. How should the doctor proceed in this case? Considering that your children have a 50% chance of carrying the allele, should they be informed? What about other family members, spouses, siblings? What if a severe and unexpected result like a Huntington's disease allele emerges in an analysis by a direct-to-consumer company? Should the company provide or facilitate the necessary medical and psychological support warranted by the severity of the case? And what happens when findings emerge within the context of a research project or clinical trial? Are researchers ethically obligated to communicate the findings and provide medical advice? The genomic data of participants in clinical trials and research projects are fundamental to genomic medicine; without genomes, there is no genomics.

Merely considering some of these possibilities and their implications, in the absence of regulations and protocols for handling them, generates the "fear of the genome" among both physicians and researchers. While still a subject of active debate, significant progress toward consensus has been made in recent years. The bioethics commission appointed by the President of the United States, after thorough analysis, suggests that in the case of genomic analyses, the type of findings mentioned are more the norm than the exception. Therefore, patients, consumers (in the case of companies offering services directly to consumers), and participants in research projects should be routinely informed about the possibility of such findings. Subsequently, depending on the type and severity of the finding, individuals should be able to choose which findings they wish to be informed about or not.

Currently, we have a widely accepted perspective as outlined by the American College of Medical Genetics, published in 2013. This recommendation provides a comprehensive list of incidental findings (specific variants) that, according to their guidelines, must be mandatorily reported if identified during genomic analysis. This detailed list includes risk allelic variants such as those associated with BRCA, among various other genetic diseases. The rationale for these recommendations, is that they are well known pathogenic variants (as discussed previously in Chap. 7 in the context of ACMG criteria), and rooted in the principle that if incidental findings are actionable—meaning there are existing therapeutic or preventive measures available—such crucial information should consistently be communicated to patients. This approach mirrors practices in other areas of medical studies where similar actionable information is routinely relayed to individuals to allow informed decision-making. In aligning with the broader considerations discussed under the ACMG criteria, this stance underscores a significant shift towards more proactive patient information strategies in genetic testing, as well as a unified framework for how and what to report.

Regarding participants in research projects and clinical trials, an important shift from traditional practices, concerns the return of genomic-based incidental findings to participants who desire them, after their subsequent validation, in a certified provider. Traditionally, in such protocols, participants contributed their samples without a specific commitment for return. For instance, in large genome projects studying population variability, samples are analyzed completely anonymously and donors are never recontacted. It is now recognized as an ethical duty to communicate significant incidental findings, along with subsequent medical assistance. Interestingly however, defining which findings to share and the costs associated with the preceding informational process and subsequent medical practice remains controversial.

Finally, a crucial unresolved issue regarding these findings is the responsibility of doctors/researchers towards third parties at risk. For example, identifying someone as a carrier of risk alleles (such as BRCA or ApoE4) or responsible for a Mendelian disease implicitly puts their children, siblings, and parents at risk. Should these individuals be informed of the risk? And under what circumstances? Incidental findings have always existed in medical practice; genomic analyses have only vastly expanded their scope, further complicating the relationship between those possessing knowledge and the individuals whose lives this knowledge affects. It is everyone's responsibility to establish an ethical framework suitable for managing this information for the common good.

Do Genes Have Owners?

If I were to ask you whether human genes have an owner, I am almost sure you would laugh at the question, considering it almost ridiculous. However, this "seemingly simple" question resulted in one of the most significant trials of the genomic era, which reached the U.S. Supreme Court, and whose ruling (June 2013) became a landmark event in the relationship between the genome and business. It all started in the late 1980s with the search for genes related to the risk of developing cancer, leading to the discovery of the previously mentioned BRCA1 and BRCA2 genes. Early on, researchers understood the potential of this knowledge as a diagnostic tool for cancer. In 1994, a team led by Mary-Claire King identified the location of BRCA1, and shortly after, scientists at Myriad Genetics isolated and sequenced the gene. That same year, the company applied for patents on the isolated BRCA1 gene and its diagnostic applications. Later, Myriad also identified and patented BRCA2. Together, these genes account for up to 5%–10% of breast and ovarian cancer cases (not 80%), and also confer an elevated risk of prostate and other cancers in men. Myriad's business model was to offer, exclusively, diagnostic analyses based on the sequences of the BRCA genes. Scientists, perhaps due to their idiosyncrasies, usually do not pay attention to such patents that could interfere with their research, and companies turn a "blind eye" as long as the discoveries are not used commercially. Moreover, acquired knowledge about patented compounds often increases their value or opens new markets. However, in the case of something as sensitive as human genes, voices of protest slowly began to rise. No company offering genetic analyses dared to conduct tests on BRCA, for fear of being sued by Myriad. This left Myriad with a monopoly on these analyses, which were widely used in cancer studies. The test case, key to advancing the trial, occurred when the Association for Molecular Pathology (AMP), along with the American Civil Liberties Union (ACLU), individual researchers, patients, and medical organizations, filed a lawsuit against Myriad Genetics and the U.S. Patent and Trademark Office (USPTO). They argued that the patents on the BRCA1 and BRCA2 genes were invalid because naturally occurring DNA cannot be patented The lawsuit against Myriad and the patent office was referred to the New York district court, which heard the following arguments. The plaintiffs argued that according to the law, patentable "any new and useful process, machine, or composition of matter, excluding products of nature," and therefore, since human genes are products of nature, they are not patentable. Furthermore, the process of comparing two genes (in this case, an individual's BRCA against a reference) and analyzing their differences (as a diagnostic method) is a "mere" mental process that simply describes a basic aspect of the scientific method. Myriad's main argument was that there was jurisprudence on the patenting of genes in the form of "isolated molecules" in a manner similar to how patents are guaranteed for other chemical compounds (for example, antibiotics derived from fungi), and that isolating the BRCA gene for analysis makes it different from the natural one embedded in the genome.

In March 2010, the district judge Robert Sweet accepted the plaintiffs' arguments and determined that genes were not patentable. Myriad appealed, and the Court of

Appeals for the Federal Circuit overturned the original ruling in 2011, siding with Myriad on the grounds that isolated DNA is chemically distinct from DNA in the body. Among the arguments used, the court relied on the judgment of the Supreme Court in a previous case about a genetically modified organism (Diamond v. Chakrabarty, 1980), where the patent was accepted on the argument that the modification makes the organism significantly different from the natural one. The court held that the gene, being isolated, is fundamentally different from the one in our body and is thus patentable. After this adverse decision, the plaintiffs appealed to the Supreme Court. Note that the arguments, for and against, focus on the very nature of the gene and our knowledge about it. Ultimately, the Supreme Court unanimously ruled in 2013 (Association for Molecular Pathology v. Myriad Genetics, Inc.) that naturally occurring DNA sequences cannot be patented.

So, what is the "important" nature of the gene? Is it its physical nature, the fact that it is within our genome and therefore, being isolated, its nature changes? Or is it its sequence, its information content, and therefore the mere act of isolating it does not represent a change in its essence at all? In June 2013, and consistent with the interpretation presented in this book, the U.S. Supreme Court held that "a naturally occurring DNA sequence is a product of nature and therefore not patentable, regardless of whether it has been isolated," and that "isolating it does not change its nature significantly in what is relevant to the law." The background of the case described is part of a broader debate that seeks to precisely delineate the difference between a discovery—which is not patentable—and an invention, which is. The sequence of a human gene is a discovery and, like the theory of gravity or electromagnetic waves, is not patentable. A synthetic gene, invented, on the contrary, can be patented, like a radio (which is based on electromagnetism).

Privacy in the Genome Era

Privacy and Data Confidentiality are also subjects of intense debate in the context of ELSI. The more radical authors contend that, in cases requiring the publication of the complete genome, it is impossible to ensure these protections. However, over the last 20 years, and as previously mentioned, an increasing number of human genomes (or parts thereof) have been sequenced and made publicly available. Current standards typically involve the de-identification of the data, provide general access only to aggregated data, and allow access to associated genotype and phenotype data strictly under institutional supervision. Although concerns about privacy and confidentiality persist, they have clearly not impeded the progress of genomic data or the availability of public genomic data.

Perhaps an intriguing way to elucidate the concerns mentioned earlier is through the following stories. When Watson, decided to have his genome sequenced and made public, he requested, for personal reasons, that information related to ApoE variants conferring risk for Alzheimer's disease be omitted. Watson publicly explained that he did not want to know or reveal his Alzheimer's risk for several reasons, including lack of Medical

Actionability, possible Psychological Burden, both for himself and his family members, as an Ethical Example, since being a public figure in genomics, Watson wanted to set an example that even the most open genomic data should still respect personal autonomy and the right not to know, and finally family Considerations, since disclosure of his ApoE status would implicitly reveal potential information about his children's genetic risks, raising a familial privacy issue. Notably, all issues that we already discussed. To comply with this request, the team of scientists in charge of the project removed the ApoE4 gene sequence from the public version of Watson's genome. A few months later, at an international conference, Dr. McGuire, responsible for managing the ethical aspects related to Watson's genome publication, was discussing these issues before a large and interested audience. At the end of the talk, one of the audience members asked to pose a question. Upon being handed the microphone and to the dismay of everyone, Michael Cariaso, an enthusiastic bioinformatics programmer, explained that by using the public data from Watson's genome, it was possible, and quite easy in fact, to infer which allelic variants of ApoE his genome possessed. He concluded with the phrase, "Watson may not know his ApoE genotype, but I do." Shortly thereafter, researchers from Australia published a paper titled "Watson's ApoE Genotype: Genetic Information is Hard to Hide," explaining how they could determine the genotype in question from public data. Beyond the embarrassment for the researchers for being outsmarted by a self-taught individual, the ethical debate highlighted by this anecdote is summarized in the following question: what exactly did Watson (or anyone making their genome public) intend to hide about his genomic information? If it was the ApoE gene sequence, this was done, but it appears to have no practical utility. If what he really wanted was not to know his ApoE genotype, here the researchers failed. Now, if what Watson really wanted was not to publish anything that could predict his risk of Alzheimer's, this demand is impossible to meet, as it is impossible to know what discoveries will appear in the future. If this story did not give you "a bit of fear," wait until you read the second one.

This is a broader case that shows why it is truly impossible to maintain the confidentiality and privacy of public genomes. In 2013, in an article in Science, a group from the Massachusetts Institute of Technology showed that by combining the information provided by the genomes with public databases available on the Internet, it was possible to determine the last name and possible place of residence, which along with additional data allowed for the identification of the genome's owner with extreme certainty. The idea is based on first determining, from the genome, certain patterns (similar to those used in paternity tests or by the police), which are then used in public databases of genealogical trees to determine possible last names and geographic area of origin. With this information plus the date of birth and other data usually associated with genomic information, it is possible, using public databases, to determine to whom the genome belongs. As a proof of concept, the researchers showed that they could positively identify Craig Venter and several participants of the 1000 Genomes Project, both Europeans and Americans.

Beyond the particular strategy used, it is evident that as we learn more and more about our genomes, and the databases of ancestry have more and more information, as the current trend shows, the possibility of keeping genomes anonymous will become increasingly difficult. In this context, it is valid to ask whether we should maintain

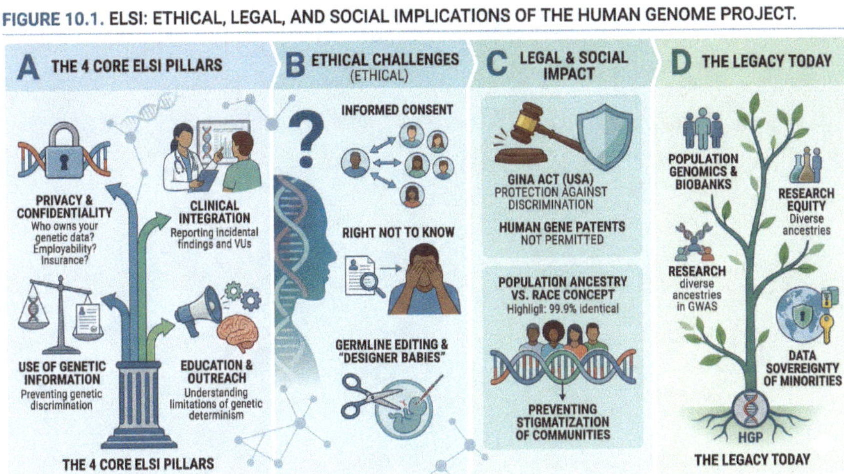

Fig. 10.1 ELSI: ethical, legal, and social implications of the human genome project. (**A**) Four core pillars, including privacy and clinical integration; (**B**) Ethical challenges surrounding consent and new technologies like editing; (**C**) Legal protections and social identity concepts; and (**D**) The current legacy of equity and sovereignty in data science

confidentiality and privacy as one of the pillars of informed consent. The foundational ethical, legal, and social pillars established by the Human Genome Project are finally detailed in Fig. 10.1, which provides the structured framework essential for the clinical integration and responsible application of genomic data.

As we conclude our exploration of the ELSI of the genomic era, it is evident that the debates around privacy, consent, and ownership are pivotal in shaping our journey into genomic science. These discussions not only illuminate the complex interplay between science and society but also underscore the crucial need to foster a balanced approach that respects individual rights while embracing the promises of genomic advancements. Embracing these challenges, we move forward, understanding that every technological advance offers both solutions and new dilemmas, forging pathways that require thoughtful consideration and responsible action in our quest to navigate the genomic frontier.

References and Suggested Further Readings

Bakshi AM. Gene patents at the Supreme Court: association for molecular pathology v. myriad genetics. J Law Biosci. 2014;1(2):183–9. https://doi.org/10.1093/jlb/lsu007.

Green RC. ACMG recommendations for reporting of incidental findings in clinical exome and genome sequencing. Genet Med. 2013;15(7):565–74. https://doi.org/10.1038/gim.2013.73.

Nyholt DR. On Jim Watson's APOE status: genetic information is hard to hide. Eur J Hum Genet. 2009;17(2):147–9. https://doi.org/10.1038/ejhg.2008.198.

Gymrek M. Identifying personal genomes by surname inference. Science. 2013;339(6117):321–4. https://doi.org/10.1126/science.1229566.

Chapter 11
Cancer Genomics

Abstract This chapter presents cancer genomics as a central domain of clinical genomics, where genomic information directly informs prevention, diagnosis, and treatment. It frames cancer as a genome-driven disease arising from somatic mutations that alter cellular behavior, leading to uncontrolled proliferation. The distinction between driver and passenger mutations is introduced as a key concept for identifying clinically relevant alterations, alongside the roles of oncogenes and tumor suppressor genes in tumor development. The chapter outlines how somatic variants are classified using oncology-specific frameworks and emphasizes the importance of assessing their clinical actionability. It further examines the complexities of tumor sequencing, including tumor purity, heterogeneity, and the distinction between germline and somatic variants. Building on these foundations, the chapter explores cancer pharmacogenomics, highlighting targeted therapies based on specific genomic alterations, as well as advanced concepts such as synthetic lethality and homologous recombination deficiency. Additional sections address oncological risk prediction through integrated genomic and clinical models, and the emerging role of circulating tumor DNA in early detection and monitoring. Overall, the chapter illustrates how genomic insights are transforming oncology into a precision-driven discipline with profound clinical impact.

Keywords Cancer genomics · Driver mutations · Oncogenes · Targeted therapy · Somatic variants

If I were to ask you what Steve Jobs (Co-founder of Apple and inventor of the iPhone) and Angelina Jolie (Hollywood megastar) have in common, few would guess the answer I have in mind: it is that both have used genomic information to fight cancer. As Isaacson narrates in his outstanding biography, Jobs used genomic information to guide the treatment of his pancreatic cancer, while Angelina Jolie underwent a preventive double mastectomy after learning she carried a pathogenic allelic variant of the BRCA1 gene, which significantly increases the risk of breast cancer. In her

M. A. Marti, *Clinical Genomics*,
https://doi.org/10.1007/978-3-032-34667-4_11

letter to The New York Times, titled "My Medical Decision" (May 2013), Jolie stated that her mother fought cancer for more than a decade and died at 56—barely enough time to hold her first grandchild in her arms. Discussions with her children about their grandmother and whether the same could happen to her ultimately led her to this decision. Her decision, amplified by social media, was perhaps the most important event highlighting the impact of the genome on our health.

Cancer is possibly the most significant disease of the twenty-first century. Justifying this statement with statistical data is almost unnecessary, as nearly all of us have lost a loved one to the disease; however, some numbers can provide a clear idea of its magnitude. It is estimated that more than seven million people currently die from cancer globally each year. Furthermore, statistics suggest that one in three women and one in two men will develop some type of cancer during their lives. In some countries, cancer will surpass heart disease as the leading cause of death. In this context, it is not surprising that one of the most anticipated promises of the Human Genome Project was its contribution to the fight against cancer. Many researchers even maintain that it is in the fight against cancer that genomics has the greatest potential to contribute clinically in the short term. This contribution also occurs on three levels: In prevention, by determining risks for various types of cancer, similar to what is seen in Chap. 9; in diagnosis, by the precise molecular characterization and classification of the tumor; and in treatment, through tumor pharmacogenomics, which acts both to assess our susceptibility to chemotherapeutic agents and to define which drugs the tumor is selectively sensitive to. However, and perhaps most importantly, cancer is in fact a disease of the genome.

Cancer as a Genome Driven Disease

Cancer encompasses a spectrum of diseases unified by the characteristic of uncontrolled proliferation, or growth, of our own cells. The accumulated mass of these cells forms what is known as a tumor. Cancer is classified based on the cell type from which it originates and its anatomical location. Thus, the unchecked multiplication of colon cells leads to colon cancer, whereas uncontrolled growth of white blood cells results in leukemia. The crucial issue is what triggers such uncontrolled cellular proliferation. The rules that govern cellular behavior are encoded in the genome, the individual's complete set of genetic material. As discussed in previous chapters, the genome, from a conceptual and biological perspective, holds every trait that defines an individual. At the cellular level, the genome offers instructions that enable a cell to survive, manufacture proteins, and carry out its functions within the body. Therefore, it is accurate to state that the genome encapsulates all the information necessary for a cell to thrive in its environment and to contribute to the overall health of the organism. Intriguingly, though all cells of a single organism possess the same genome, they can differentiate into diverse types such as hepatocytes, neurons, or erythrocytes (red blood cells). What differs, if the genome remains unchanged, is gene expression—specifically, which genes are activated or repressed in each cell type. This gene expression is influenced by the surrounding environment or the context in which the cell exists. In summary, the genome acts as a set of instructions that interacts with its environment to ensure that the cell functions harmoniously within the organism. This concept helps explain why cancer is often described as a genomic disease. When the genome of a cell is altered, its instructions are modified, causing the cell to deviate from its specific functions—such as those performed by a normal hepatocyte or neuron—and instead begins to operate autonomously, guided by its own 'internal desire'. This shift in cellular behavior, driven by genetic changes, results in the cell losing control over its growth and division. Consequently, it acquires tumoral characteristics and forms a tumor. Thus, we assert that cancer is fundamentally a genomic disease: its root cause stems from alterations in the genetic instructions that force a cell to disregard the well-being of the organism and trigger, internally, a detrimental effect that may even lead to the death of the individual. Each cancer then begins with a cell that, due to alterations in its genome, starts to divide, proliferate, and invade its surroundings. Similar to what was seen in a previous chapter regarding the polymerase chain reaction, let us imagine that if a human cell is capable of duplicating every 24 h, a single tumor cell can, over the course of a month, give rise to over a million cancerous cells. During this progression and exponential growth, the cells accumulate more and more changes in their genome, mutations that make them more aggressive and difficult to control. Hence, one of the keys to defeating cancer is its early detection. Let's deep dive now into the nature and characteristics of these mutations.

Driver and Passenger Mutations

The modern understanding of cancer as a genomic disease is relatively recent, having developed over the last century alongside the history of molecular biology. A pivotal milestone occurred at the start of the twentieth century, when Peyton Rous demonstrated that a virus—the Rous sarcoma virus—could induce tumor development in birds. This finding led to the hypothesis that cancer could be an infectious disease, an idea that aligned with the scientific mindset of the time. With the advancement of molecular biology in the 1960s, it was discovered that viruses contain genes, and that these genes could alter the behavior of infected cells. From there, focus shifted to genes as the agents responsible for cancer, rather than the viruses themselves. By the 1980s, it was confirmed that normal human cells contain genes in their own genomes capable of inducing cancer: the so-called proto-oncogenes, which, when mutated, become oncogenes. Notable among these are RAS and MYC, whose identification marked a turning point in the understanding of the molecular nature of cancer. Concurrently with these developments, another line of thought emerged that viewed cancer as a hereditary genetic disease, beyond the role of oncogenes. In this context, the idea of cancer as a recessive disease was proposed, formulated in Dr. Alfred Knudson's two-hit theory, based on his study of patients with hereditary retinoblastoma. Knudson noticed that in affected families, there was a clear hereditary transmission of cancer risk. By analyzing family patterns and applying Mendel's laws, he concluded that patients inherited one defective copy (allele) of the responsible gene. This implied that only a second mutation, acquired randomly during cell division or replication, was necessary to produce the tumor. Thus, in families carrying the mutated allele, tumors—such as ocular retinoblastoma—appeared with high frequency and at early ages. Conversely, in sporadic cases, where both alleles were normal at the start, two independent mutations, one in each allele, were required for the disease to manifest. From this theory emerged the concept of tumor suppressor genes: genes whose normal function is to prevent the cell from becoming cancerous, such as Rb, TP53 and BRCA1 and BRCA2. When both copies of these genes fail, the cell loses its regulatory capacity and can transform into a cancer cell. In the twenty-first century, with the advent of the Human Genome Project and massive sequencing technologies, thousands of tumors began to be sequenced, and the genomic alterations characterizing them were analyzed in detail, thereby consolidating the modern view of cancer as a disease of the genome.

As discussed in earlier chapters, each person carries a vast array of genetic variants, accounting for a substantial part of human genetic diversity. The variants inherited from our parents are known as germline variants because they are transmitted through germ cells—eggs and sperm. When we refer to inheriting "half of the genes from dad and half from mom," we are essentially saying that we inherit half of their genetic variants. Beyond these inherited variants, new variants may also develop during embryonic or postnatal stages. These arise not from parental transmission but, de novo, due to spontaneous or induced errors in DNA replication. It is crucial to remember, at this point, that cancer originates from a cell that was initially normal—a cell whose DNA aligns with the individual's genome but later acquires certain new mutations. These variants, which occur in non-heritable cells, are termed somatic variants. They have the potential to alter the genome's instructions and transform a healthy cell into a cancerous one. Significantly, these somatic variants are not transmitted by subsequent generations.

Among all the somatic variants that a cell or tumor acquires, not all are relevant in relation to tumorigenic capacity, or clinically relevant in an oncogenic context. Those that do not alter cellular function and can be located outside or within genes relevant to cancer are called passenger mutations. In contrast, the mutations known as driver mutations functionally impact the cell, turning it into a cancerous cell. Therefore, the goal in the genomic characterization of a tumor is to identify the driver mutations among the multitude of variants present. A tumor may contain thousands of mutations, but only a few are truly relevant to its behavior. Hence, the challenge—and the core of genomic work in oncology—is to distinguish which mutations are drivers and understand how to use them to guide diagnosis, prognosis, and treatment of cancer. This distinction between driver and passenger variants can be understood by analogy with the genetics of hereditary diseases discussed earlier: driver mutations are akin to pathogenic variants, while passenger mutations resemble benign variants. Similarly, as in Mendelian diseases where genomic or molecular diagnosis, which involves determining the pathogenic variant causing the pathology, allows for an extremely precise characterization of the pathology, the determination of the driver variants, which are causal to the tumor, enables a precise genomic or molecular diagnosis of the tumor itself.

To conclude this section, and in direct relation to tumor genomic characterization, we can delve into an important question: how many and what types of mutations does a tumor present? The answer, as expected, is that each tumor is unique, just as each individual is, but nonetheless, certain clear trends have emerged. These trends result from the ultimate cause of the tumor, namely the origin of its driver mutations. Some tumors originate from interactions with mutagenic environmental agents, known as carcinogens. For example, cigarette smoke contains multiple compounds that damage DNA; ultraviolet radiation causes characteristic mutations in skin cells. In these cases, exposure to carcinogens increases the frequency of somatic mutations, thus "accelerating" the emergence of genetic errors. Generally, the greater the exposure, the higher the mutational load of the resulting tumor. On the other hand, there are intrinsic causes. If one of the first mutations occurs in a gene responsible for maintaining genomic stability or repairing DNA, the cell loses some of its capacity to control errors in future cell divisions. This leads to an uncontrollable increase in mutations, which contributes to the genetic heterogeneity of the tumor. Mutations also accumulate with age. Our cells continuously divide to renew tissues. Each time a cell replicates, there's a chance—albeit small—that new mutations will occur. Therefore, when comparing tissues from a newborn to those of an older individual, it is evident that older individuals accumulate more somatic mutations. This accumulation also depends on the type of tissue: tissues with a high rate of cellular renewal (such as skin, intestines, or bone marrow) tend to accumulate more mutations than those with low cellular turnover. Thus, tumors can present enormous differences in the quantity and type of mutations. This diversity poses a significant challenge: how do we classify and organize tumors when each has such a distinct genomic profile? This question—which underscores the need for genomic characterization of tumors—is the starting point for modern diagnostic and treatment strategies in molecular oncology.

Oncogenes and Tumor Suppressor Genes

When studying driver mutations, it's crucial to understand that not all act in the same way. These mutations can appear in two major categories of cancer-related genes: oncogenes and tumor suppressor genes. Both contribute to tumor development when altered, but they do so through virtually opposite mechanisms.

Oncogenes are genes that, when mutated, actively promote cellular proliferation. Under normal conditions, their non-mutated versions—proto-oncogenes—participate in necessary physiological functions, such as controlled growth, intracellular signaling, and response to external stimuli. However, a gain-of-function mutation can convert them into constitutively active oncogenes, driving the cell to divide uncontrollably. These mutations often behave dominantly: a single mutated gene copy is enough to alter the cellular phenotype. Furthermore, they tend to concentrate in very specific regions of the gene or protein, as they must modify critical points that regulate their activity. Classic examples include EGFR (which we will discuss later), KRAS, BRAF, PIK3CA, and MYC, all with very confined recurrent mutations such as KRAS G12D or BRAF V600E. It is currently estimated that there are between 80 and 120 well-established oncogenes with solid evidence of their driving role in cancer.

Conversely, tumor suppressor genes act as brakes on cellular growth. Their function is to restrict proliferation, repair damaged DNA, induce apoptosis when appropriate, or maintain genomic stability. When these genes undergo loss-of-function mutations, the cell loses the ability to control its growth and survive adequately. This type of mutation usually manifests as a recessive trait: both gene copies must be inactivated for the protective function to disappear completely, which explains Knudson's classic "two-hit" theory. Representative examples include TP53, the most frequently mutated suppressor gene in cancer; RB1, which controls the cell cycle; BRCA1 and BRCA2, critical in homologous recombination repair; and APC, essential in the Wnt pathway signaling of the colon. It is estimated that there are between 200 and 300 known tumor suppressor genes, most vulnerable to various types of alterations, from point mutations to complete deletions.

In summary, as illustrated in Fig. 11.1, cellular homeostasis depends on a balance between signals that stimulate growth (oncogenes) and signals that restrict it (tumor suppressor genes). The development of a tumor occurs when this balance is systematically broken in favor of uncontrolled proliferation. In this sense, oncogenes and tumor suppressor genes represent the two opposite poles of the same process: the former dangerously accelerates the cell cycle, while the latter loses its ability to brake it.

Classification of Somatic Variants

The classification of somatic variants employs a methodology derived from the ACMG framework previously discussed in relation to germline variants. However, it features specific modifications tailored to the field of oncology. Similarly to the germline

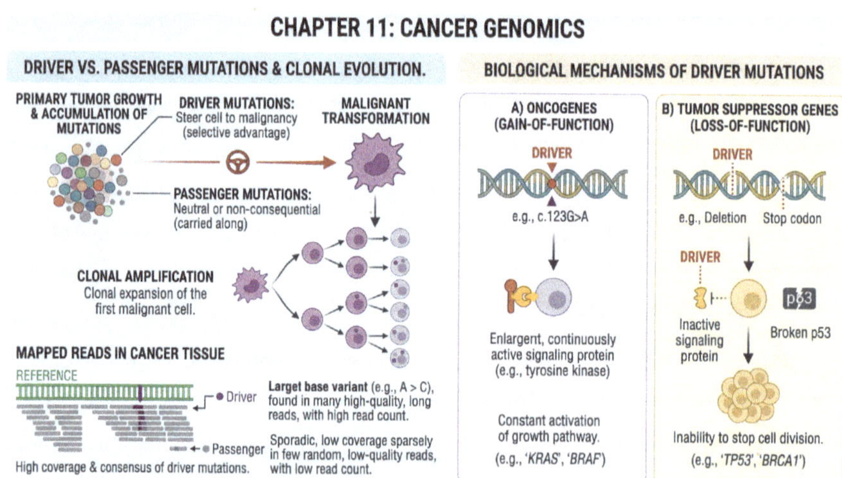

Fig. 11.1 Cancer genomics. Driver vs. Passenger mutations. Clonal amplification. Oncogenes and tumor suppressor genes

approach, it maintains five categories: Benign, Likely Benign, VUS (Variants of Uncertain Significance), Likely Oncogenic, and Oncogenic. Notably, the terminology shifts from "pathogenic"—typically used in the context of diseases caused by the variants—to "oncogenic". This differentiation is crucial as it highlights the role these variants play as primary drivers in the transition of cells to a tumorigenic state; (Likely) Oncogenic variants are identified as the key mutations leading to cancer.

In adapting the ACMG system to oncology, the complex alphanumeric code system (such as PM1, PS3, PP2, etc.) used for germline variants is replaced by a straightforward direct scoring method. This method assigns positive scores for oncogenic evidence: Supporting (+1), Moderate (+2), Strong (+4), Very Strong (+8), and equivalent negative scores for evidence supporting benign interpretations. All scores associated with variant annotations are then aggregated, and the summed score determines the final classification of the variant.

The consolidated scoring range stratifies the variants as follows: Total Score: ≤ -7, classified as Benign Total Score: –6 to –1, classified as Probably Benign Total Score: 0 to 5, classified as VUS Total Score: 6–9, classified as Likely Oncogenic Total Score: ≥ 10, classified as Oncogenic

For instance, reaching an Oncogenic classification typically requires combining scores such as one Very Strong (8) and one Strong (4) label, achieving a total of 12 points. This method mirrors the logic of the germline ACMG system but eliminates the reliance on arbitrary combination rules. The use of a numerical scoring system also means that potential contradictory evidence is directly balanced, minimizing the occurrence of Variants of Uncertain Significance due to conflicting labels.

As for the types of evidence used in scoring variants, they mirror those utilized for germline variants. The first type concerns population frequency: a high population frequency >5% yields a Very Strong benign score (–8), a moderate frequency >1% contributes to a likely benign classification, and low frequency enhances the probability of oncogenicity, scored as Supporting (+1). The second type, functional evidence, parallels the ACMG germline's criteria PS3/BS3, which rely on functional studies of genes and/or proteins. Molecular evidence is also considered, particularly regarding the type of mutation and its anticipated effect, such as nonsense mutations, frameshifts, exon loss, or canonical splicing site mutations. It's important to highlight that in oncology, many variants have been pre-classified, which considerably aids their annotation and subsequent categorization in clinical settings.

It is crucial to understand that the classification of a variant does not automatically warrant its reporting. Similar to the processes observed in germline variant classification, merely identifying a variant as (likely) oncogenic is insufficient for clinical reporting. For a somatic variant to be reported, it must meet two critical conditions: it must be (likely) oncogenic and clinically relevant, which, in the context of somatic variants, translates to demonstrating actionable potential. The system for classifying somatic variants includes a secondary category system that ranks variants into one of four clinical relevance Tiers. This tiered system does not re-assess the inherent quality of the variant—as this is established during oncogenicity classification—but it evaluates its real-world clinical impact, particularly in how it relates to existing therapeutic options.

Tier 1 corresponds to variants of High or Strong Clinical Significance and is subdivided into Levels A and B. Level A identifies variants that are associated with an FDA-approved drug and for which there are specific clinical guidelines recommending treatment. An exemplary case are certain EGFR mutations for which treatment with erlotinib is advised. On the other hand, Level B includes variants for which robust clinical studies, conducted by expert researchers and having significant statistical power, have suggested therapeutic efficacy but have not yet received FDA approval. These studies are frequently published in prestigious medical journals like The Lancet.

Tier 2 pertains to variants with Potential Clinical Relevance, akin to the concept of "likely pathogenic" but focused on clinical implications rather than mere pathogenicity. This tier includes Level C, where there is indirect or extrapolated evidence (e.g., an FDA-approved drug for the variant exists but in a different kind of tumor), and Level D, where preliminary evidence of a therapeutic benefit is available from sources like preclinical trials, case reports, and small series that lack robust statistical grounding. Variants falling under Level D are typically found in contexts involving rare or pediatric tumors, where assembling sufficient cohorts for formal trials proves challenging.

Tier 3 relates to Variants of Uncertain Clinical Significance. These include potentially oncogenic variants that lack sufficient evidence linking them directly to cancer. Commonly, they are variants found in new or poorly studied genes, leaving them in a diagnostic "grey zone." Although these variants are not disregarded, their clinical utility remains ambiguous.

Tier 4, finally, is assigned to variants with No Clinical Relevance, such as those that are (likely) benign.

In essence, while the classification of somatic variants mirrors that of germline variants used for diagnosing rare diseases per the ACMG guidelines, it necessitates an additional step. This subsequent phase involves evaluating the clinical relevance of variants post-oncogenicity assessment, with a particular focus on their actionable potential. Since actionable results are pivotal in cancer treatment and management, understanding the nuances between germline and somatic variants is essential for advancing precision oncology.

Having established, the genomic nature of the tumors and the role played by the variants, we now provide some insight into the peculiarities concerning sequencing methods for tumor samples, and their bioinformatics processing.

Navigating the Complexities of Tumor Genomics and Variant Analysis

Before initiating somatic genomic analysis, it is essential to evaluate multiple factors related to the sample, the type of tissue, and the sequencing methodology. These factors directly impact the quality of the data and the subsequent ability to detect relevant variants. In a germline DNA sequencing protocol, DNA is extracted from the patient's normal cells (from blood or saliva) and their genome is analyzed. However, in somatic sequencing, the priority is to ensure that the DNA analyzed actually comes from true tumor cells. To study the tumor's genome, the sample must contain the highest possible proportion of neoplastic cells. The identification of these cells depends on the pathologist, who selects the tumor region from the biopsy. In this context, we define tumor purity as the proportion of tumor cells relative to normal cells in the surrounding tissue. Lower tumor purity has a direct impact on the analysis, as the somatic variants will be underrepresented in the readings.

A second aspect to consider is that formalin-fixed paraffin-embedded (FFPE) tumor tissue is the standard sample in most health care settings around the world. However, this type of preservation introduces two essential problems for the study of the tumor genome: DNA degradation (small, unstable fragments are obtained), and specific fixation-induced modifications such as cytosine deamination (C→T) that can confuse variant calling. Therefore, concerning the library preparation technique, it is important to highlight that capture methods such as those used for exome sequencing often fail in FFPE samples due to the reduced size and poor quality of DNA. For this reason, methods based on amplicons, such as multiplexed PCR, tolerate degraded DNA better and are generally preferred.

It is also important to highlight the role of tumor heterogeneity. It is crucial to understand that a tumor is not a uniform mass of identical cells, but a heterogeneous population. Each tumor cell has divided and accumulated different mutations over time. Therefore, within the same tumor, there are cellular subpopulations (clones) that differ from each other by the driver mutations they possess. Thus, a genetic mosaic is formed where each subclone presents a unique combination of driver and passenger mutations. This clonal heterogeneity explains why tumors are dynamic, evolve over time, and respond differently to treatments. Each subpopulation may have different proliferative, invasive, or therapeutic resistance capabilities, depending on their particular driving mutations. These variations impact bioinformatics analysis and the quality and quantity of variants detected, and therefore must be considered before sequencing to ensure the legitimacy of the somatic analysis.

Turning our attention to variant calling, it is important to understand that since tumor cells originate from an individual's cells, they carry all the same variants, which are referred to as germline variants. On the other hand, the tumor possesses the somatic variants that arise only in the tumor during its evolution, which will be present in different proportions, reflecting the population of the subclone that contains them. Therefore, after performing sequencing, mapping, and marking the differences between the bases observed in the readings and those in the reference genome, we must analyze not only if this is a sequencing error or a germline variant but also evaluate if it is a somatic variant, and estimate, according to its read depth, in which fraction of the tumor cells it is present. In practice, when performing variant calling on a tumor sample, the process will first involve calling variants, which does not classify them as homozygous or heterozygous according to their read depth, and will even call those variants supported by low depth. Thresholds of 5% and five to ten reads are usually used in 100X sequencing, allowing us to maintain the depth and determine a lower clonal percentage for theoretical higher depths. Once this initial variant calling is completed, the task is to distinguish which are germline and which are truly somatic.

Distinction of germline and somatic variants presents two possible scenarios. The first and optimal scenario involves having sequenced both the tumor cells and the adjacent normal tissue cells. This scheme, known as paired analysis, allows for the assignment of variants called in the normal sample as germline, leaving those observed exclusively in the tumor as somatic. Although this scenario is ideal, it rarely occurs outside of research settings because it requires sequencing two samples, resulting in double the cost. In a clinical genomics context, it is more common that only the tumor sample and its sequencing are available, so after variant calling, variants are assigned as germline based on their probability of occurrence in the population as determined by population frequency. It is important to note that while this method may be imprecise, assigning variants as somatic that are in reality germline, the crucial aspect is to determine if a variant is a driver, that is, oncogenic, and then if it has clinical relevance. The variant will be relevant whether it is germline or somatic. Furthermore, once the variant is identified, even if it is germline, it will be present in the tumor and thus allow for appropriate therapeutic measures. Subsequently, its somatic or germline nature can be verified by targeted sequencing of the same in a normal sample. In other words, classifying a variant as germline does NOT mean discarding it, but rather placing it in the correct context.

Principles of Cancer Pharmacogenomics

Understanding that cancer arises from driver mutations—and distinguishing between mutations in proto-oncogenes and tumor suppressor genes—first opened the door to truly targeted therapies, giving rise to the concept of precision medicine. If a proto-oncogene is activated by a driver mutation, this activation occurs exclusively in the tumor cells. In healthy tissues, that same gene is not mutated and, moreover, is not usually active continuously. This means that identifying a driver mutation in a proto-oncogene automatically offers a molecular target that is only active in the tumor. Targeting that protein therefore allows for specific treatment of the malignant cells without significantly harming the rest of the body. Similarly, when the driver mutation appears in a tumor suppressor gene, although this generally implies a loss of function, the end result can also be the aberrant activation of a cell proliferation pathway. In both scenarios, understanding the mutation allows identifying cellular pathways that are active exclusively in the tumor and, consequently, inhibiting them through drugs designed to block that pathological signaling. Thus, the so-called targeted therapies were born, which are the core of the modern concept of precision medicine in oncology and the main concept of cancer pharmacogenomics.

This approach contrasts strongly with classic chemotherapy, where, not knowing the genomics of the tumor, the only thing that can be asserted with certainty is that cancer cells divide more rapidly than most normal cells. This difference in division rate was, historically, the basis for the first oncological treatments. The treatments known as traditional chemotherapy attacked all dividing cells indiscriminately—including tumors, but also hair, gastrointestinal mucosa, and bone marrow—generating characteristic adverse effects: hair loss, digestive problems, anemia. These were, or are, generic treatments, useful but not very specific. The radical change occurred in the early years of the twenty-first century, when our understanding of driver mutations began to guide specific therapeutic decisions.

It is important to note, that the therapeutic utility of tumor genomics is not limited to inhibiting proliferative pathways. As previously described, many driver mutations involve failures in DNA repair mechanisms. The tumor's inability to repair genetic damage leaves it particularly vulnerable to certain drugs that force it to rely on alternative repair pathways, thus exploiting a phenomenon known as synthetic lethality, which we will explore further later. Another key consequence is that mutations generate altered proteins—neoantigens—that differ from the body's normal proteins. From an immunological standpoint, this turns the tumor into an "alien organism," allowing the immune system to be stimulated to recognize and eliminate it, which is the context for the tremendous development of immunotherapy to treat cancer.

To advance our understanding of cancer pharmacogenomics and how it leads to the concept of precision medicine, we will briefly analyze one of the historical and paradigmatic examples of this, relating to lung cancer. In the early 2000s, studying the genomics of these tumors, it was discovered that a significant fraction of them exhibited mutations in a key protein in cellular growth signaling: the epidermal growth factor receptor (EGFR). This discovery marked the beginning of molecular oncology as we know it today.

EGFR (Epidermal Growth Factor Receptor) functions normally as a growth factor in epidermal cells by responding to external signals that indicate when to activate and promote cellular replication. However, when an activating mutation occurs, the receptor may turn on autonomously—even without a signal. This spontaneous activation enables cells to proliferate uncontrollably, a phenomenon that is extremely prevalent in tumors. EGFR is integral to a broader signaling pathway, and the development of inhibitors targeting this pathway has become a critical therapeutic strategy. If a tumor exhibits a specific mutation in EGFR, it is feasible to pharmacologically inhibit its kinase activity, thereby reversing uncontrolled proliferation.

EGFR is a complex protein comprising multiple domains, articulated through various modules along its sequence. It operates as a dimer, where each monomer binds to the other upon receiving the signal from its ligand. Dimerization activates the kinase at its cytoplasmic end, which then phosphorylates other proteins, such as AKT and PI3K—both of which are critical kinases in cellular signaling. These events trigger a signaling cascade that leads to cellular proliferation. In essence, any variant that increases the likelihood of EGFR activation without an external signal categorizes the gene as a tumor driver. EGFR thus epitomizes a classic proto-oncogene that, upon mutation, acquires an exaggerated function. The mutations responsible for EGFR hyperactivation have been extensively identified and cataloged, so currently, it is not necessary to predict de novo whether a variant will be activating; detection suffices. When a tumor bears one of these recognized mutations, a specific inhibitor can be accurately selected to obstruct cellular growth. This knowledge is why mutations in EGFR are routinely examined today in the management of lung cancer—especially in the non-small cell lung cancer (NSCLC) subtype. Upon identification of one of these mutations, the selection of first-generation tyrosine kinase inhibitors such as erlotinib or gefitinib can commence as the first-line therapeutic approach. These inhibitors are designed to selectively deactivate the hyperactive pathway.

This set of observations consolidated several fundamental concepts of precision medicine: the identification of tumor drivers, the activating nature of mutated proto-oncogenes and the use of targeted therapies. From these advancements, the field has grown exponentially. A few years ago, it became clear that to treat a tumor properly, it was essential to know its mutations to select the initial treatment and, moreover, to systematically monitor its molecular evolution. This led many physicians to begin routinely requesting genomic studies.

Synthetic Lethality and Homologous Recombination Deficiency (HRD)

The cell employs various specialized mechanisms to maintain the integrity of its DNA. Some of these mechanisms act by correcting errors that appear during replication, such as base excision repair (BER), which removes and replaces chemically damaged bases, and mismatch repair (MMR), which is responsible for recognizing and correcting base mismatches or small insertions/deletions generated during DNA synthesis. Other mechanisms handle more severe injuries, such as double-strand breaks. Among these, non-homologous end joining (NHEJ) reconnects the broken ends directly and swiftly, albeit with limited accuracy that may introduce mutations, including significant loss of genetic material. In contrast, homologous recombination (HR) uses the intact copy of the homologous chromosome to restore the original sequence, thus providing the most faithful and conservative pathway for repairing this type of damage.

The proteins BRCA1 and BRCA2, along with other genes that frequently harbor driver mutations in cancer, are essential for the proper functioning of homologous recombination. Although BRCA1 and BRCA2 were identified several decades ago and that their full range of functions has not been completely clarified, it is known that they participate in a sequential and highly coordinated manner in the repair of double-strand breaks. BRCA1 acts in the early stages, regulating the access of various protein complexes to the damage site and organizing the architecture of the repair machinery, promoting the controlled resection of the DNA ends. BRCA2 intervenes in later steps and plays a central role in loading and stabilizing RAD51, a process facilitated by the protein PALB2, thus enabling the formation of the nucleoprotein filament required for homology search and precise DNA repair.

While HR is involved in the precise repair of double-strand breaks, there are other complementary pathways responsible for resolving less complex damage, such as single-strand breaks. In this context, poly-ADP-ribose polymerases (PARP) play a central role. PARP1—the most relevant isoform in cancer—recognizes various types of DNA lesions rapidly and efficiently via zinc-coordinating domains, similar to those found in transcription factors. Once associated with the damage site, PARP1 uses ADP-ribose as a substrate to synthesize poly-ADP-ribose chains on itself, as well as on other nearby proteins. These modifications act as signals that recruit and organize the protein complexes responsible for DNA repair, facilitating mechanisms and processes that resolve single-strand breaks, thus contributing to the overall maintenance of genomic integrity.

During the 1990s the first PARP inhibitors were developed as research tools to study DNA repair. These initial compounds had no therapeutic purpose; their utility was primarily experimental. However, around 2005, a turning point occurred: two independent groups formulated and demonstrated a key hypothesis. They proposed that since cells deficient in BRCA1 or BRCA2 lack functional homologous recombination, they should become extremely vulnerable if single-strand break repair is simultaneously blocked, for instance, through PARP inhibition. By testing this idea, they observed that PARP inhibitors were selectively cytotoxic in cells with BRCA1/2 mutations, while normal cells tolerated the treatment. This finding constituted the experimental demonstration of the concept of synthetic lethality in cancer and paved the way for the clinical development of PARP inhibitors as a targeted strategy for tumors deficient in homologous recombination.

To understand why certain tumors with BRCA alterations are sensitive to PARP inhibitors, it is necessary to delve deeper into and grasp the concept of synthetic lethality. This term describes a situation where a cell can survive the failure of one of two essential systems (A or B), but dies if both fail simultaneously. In this context, the two critical systems are the homologous recombination repair pathway and the single-strand break repair pathway mediated by PARP. A tumor cell deficient in homologous recombination can survive, albeit by generating mutations derived from imperfect repairs. However, if PARP is also inhibited—blocking the repair of single-strand breaks—the cell loses both fundamental mechanisms and accumulates lethal DNA damage. When single-strand errors are not repaired, they evolve into double-strand breaks. If homologous recombination is compromised, these breaks cannot be corrected properly, and the cell ultimately dies. This is the central principle behind the therapeutic use of PARP inhibitors in tumors with BRCA defects or other genes in the same functional circuit.

Therefore, to utilize PARP inhibitors as a therapy, it is crucial to determine which tumors are deficient in the mechanism of homologous recombination. The direct way to demonstrate this deficiency is by identifying the presence of loss-of-function (oncogenic) variants in genes such as BRCA1, BRCA2, RAD51, and PALB2, among others. In this context, the following dilemma arises in the clinical practice: Imagine a physician orders a study using a small panel of genes associated with HR and finds no driver mutations. In this scenario, the decision to prescribe or not prescribe a PARP inhibitor must be made. From the pharmaceutical industry's perspective, the recommendation is often to offer it anyway, arguing that HRD may exist even without detectable mutations in BRCA. On the other hand, health insurance companies or government-based systems, usually demand stricter evidence before authorizing this type of treatment. This tension between clinical, regulatory, and economic criteria constitutes one of the most intense debates today, and this is where clinical genomics provides assistance. The fact is, that even in the absence of oncogenic mutations in HR-associated genes, the tumor may still be deficient, and mass sequencing technologies can identify it.

Tumors deficient in homologous recombination exhibit characteristic patterns of genetic instability: deletions, loss of material, and mutations derived from NHEJ-type repairs. These patterns can be identified, for example, by detailed analysis of molecular signatures associated with somatic variants, based on whole-genome sequencing. A more economical variant involves the use of gene panels that evaluate the status of Homologous Recombination Deficiency status (HRD). These panels analyze beyond the presence of driver mutations in the HR-associated genes; three types of mutations characteristic of HRD, which result in their resolution by NHEJ. These include, firstly, loss of heterozygosity (LOH). This occurs when a region of the genome is duplicated and both copies become identical, eliminating allelic variability, or when a complete allele is lost, leaving only one copy (i.e., a large deletion appears). In both cases, the region ceases to be heterozygous and becomes homozygous or hemizygous. The second component is telomeric allelic imbalance (TAI), which corresponds to a deletion that begins at any point on the chromosome and extends to the end of the telomere. Although related to LOH, it adds the requirement that the alteration necessarily extends to the end of the chromosome. Since there are 23 chromosomes, each with 2 arms, the maximum number of TAI instances is 46. The third, and last, component encompasses changes in copy number variations or translocations that cover regions larger than 10 megabases. Notably, these three types of variants result from the fusion of chromosomal ends with the consequent loss of genetic material, characteristic of the failure of the homologous recombination system. By counting the total number of these three types of alterations, an HRD score is obtained, which indirectly evaluates the accumulated history of double-strand breaks in the tumor genome and therefore the HRD status. Tumors with high scores, as expected, have been shown to be more sensitive to the use of PARP inhibitors.

As the understanding of tumor biology advances, the use of PARP inhibitors guided by precise genomic diagnostics exemplifies personalized medicine's potential. This approach enhances treatment efficacy and minimizes side effects for patients with specific genetic profiles. Therefore, advancing these diagnostic techniques is critical for improving cancer treatment outcomes. Having analyzed the genomic nature of the tumor, and the impacts in the underlying tumorigenic mechanisms, and potential treatments in the context of cancer pharmacogenomics, we now turn our attention back to the concepts analyzed previously in Chap. 9 and integrate them into current models of oncological risk prediction.

Oncological Risk Prediction

The primary objective of risk estimation —in oncology or any other disease as we have seen in previous chapters— is to calculate the lifetime risk. This value represents the percentage of individuals in a group with certain characteristics—in this instance, both genetic and clinical—who will develop the disease at some point in their lives. For example, if a model assigns a 15% risk to a given group, it means that out of every 100 people with that profile, 15 will likely develop the pathology. Every disease has a baseline risk, representing the risk observed in the general population without particular characteristics. From this value, estimated risks for different population subgroups are compared. In most tumors, we are not dealing with extremely high risks; the values tend to be below 40%. Therefore, populations are usually stratified into three categories: population risk, moderate risk, and high risk, and this stratification allows for the adjustment of preventive strategies. In breast cancer, for example, the average population risk, which is below 15%, results in the initiation of biennial mammography starting at age 50. However, those women with a moderate risk (in the 15%–30% range) may require an earlier start of screening, the addition of more sensitive studies such as breast MRI, and addressing modifiable factors. In the group displaying a high risk (>30%), the onset of screenings is brought forward even more, surveillance is intensified, and chemical or surgical prevention strategies may be considered. Thus, the goal of genomics is to determine whether genetics keeps a person within the population risk or shifts that risk to the moderate or high categories.

In breast cancer, the lifetime risk comes from several factors with underlying different biological mechanisms. There are high-penetration genes—such as BRCA1, BRCA, or TP53—whose oncogenic variants significantly increase the risk of developing cancer. These mutations are often driver variants that facilitate the emergence of the tumor by reducing the need to acquire additional events. For this reason, a person carrying these variants has a substantially higher risk than someone from the general population. However, in addition to these high-impact variants, there is also a much larger set of low-penetration variants, which behave in a polygenic risk model. Each contributes modestly, but collectively—being distributed across many genes—they produce a significant cumulative effect. The variants used in these models, as we have previously seen, are those discovered in GWAS, and their impact is measured by their corresponding Odds Ratio (OR). Finally, in oncological diseases, the environment and life choices also hold a tremendous potential contribution to the lifetime risk, as evidenced by the associations between cigarette smoke and lung cancer, or exposure to the sun's ultraviolet light and melanoma. For breast cancer, the greatest contributions are provided by the onset (menarche) and end of menstrual periods, the number of children, and body mass index, among others.

One of the earliest models to recognize the multifactorial nature of the risk of developing breast cancer was developed by Tyrer-Cuzick at the beginning of the 21st century, and is widely used in the field of breast cancer. This was the first model to integrate genomic risk with several other factors and, in particular, included two high-penetration genes—BRCA1 and BRCA2—along with "all other genes," grouped as a single polygenic component. In other words, it made an initial approximation to a polygenic risk model, which was inferred from an analysis of family history. If an individual did not have mutations in BRCA1 or BRCA2 but had a strong family recurrence, the model interpreted this recurrence as the phenotypic manifestation of variants distributed at other loci. With this information, it adjusted the individual risk. This model—which has been refined over the years with large cohorts—allows the estimation of curves of accumulated age-related risk for the general population and for carriers of driver mutations in BRCA1/2. In the last 20 years, with the aforementioned emergence of GWAS, the polygenic component began to be determined with greater precision, and currently, risk models include contributions from between 300 and350 genes.

In the present day, the most developed model is called BOADICEA and seeks a comprehensive approach to predicting the risk of breast cancer integrating multiple dimensions of risk, including polygenic risk based on hundreds of genetic variants, non-genetic factors, clinical elements, and lifestyle characteristics. It even has an online tool where patient data is entered and the system returns personalized risk estimates. The model determines for each main component, an adjusted OR that multiplies the initial (a priori) risk and then adjusts the lifetime risk, and the risk of developing cancer in the next 10 years, for each patient according to their characteristics. The model classifies the risk as low, moderate, or high. Notably, a high risk implies more than a 30% cumulative, or more than an 8% to 10-year risk. Analyzing the contribution of genetic components, it is evidenced that oncogenic mutations in high-penetration genes can contribute with an OR of up to 4, a value similar to that obtained for individuals in the upper percentile of the polygenic risk model. It is interesting to note that, while genetic factors are fixed and do not change throughout an individual's life, other factors related to life choices do. For example, one of the main effects on risk is the number of children. Women without children have a baseline risk, but with one pregnancy the risk decreases slightly, and with two or three pregnancies it decreases even more, achieving reductions of up to 30%. The Body Mass Index, although with a lesser effect, also helps to reduce the risk.

From a population's perspective, the goal of determining each woman's risk is to stratify them in order to assign interventions rationally. Those with low risk may require fewer studies or interventions; those at average risk follow standard recommendations; those at high risk require intensive surveillance or more active preventive measures. From the patient's perspective, it is crucial to explain and understand the following key aspects of the results. First, that these models always provide probabilities, not certainties. Even at a very high risk, a patient may not develop cancer. This is particularly true if the high risk assignment is used to perform preventive measures. Second, that the fact the risk is probabilistic does not mean it is inaccurate. In clinical practice, the procedure is always the same. A patient arrives, their genotype is determined for the variants of the model, their score is calculated, and then the individual is placed within the distribution. If they fall into the high-risk group, intensive surveillance, preventive measures, or specific interventions are recommended. If they fall at the opposite end, strategies appropriate to the low risk are adopted. The logic is to stratify and act according to the level of individual risk, not the population average.

Finally, it is important to highlight that different polygenic and clinical risk models for other types of cancer are under active development. The closest example is prostate cancer, for which there are well-developed polygenic scores, but integrative models are not yet available. Research continues to advance, but models for other tumors are still in a less mature stage.

Tumor Cell-Free DNA (cfDNA) Analysis

In this final section, we turn to Tumor Cell-Free DNA (cfDNA) Analysis, a rapidly evolving area that is transforming how clinicians and researchers detect, track, and characterize cancers using liquid biopsies. One of the long-standing aspirations—often described as the "holy grail" of early cancer diagnosis—is the ability to detect a tumor before it becomes visible on imaging or before its biological effects produce clinically recognizable symptoms. Achieving this for previously unknown primary tumors remains extremely challenging, yet significant progress has been made in the early detection of metastatic disease in patients who had already undergone surgical and/or systemic treatment.

The general principle is as follows: if a patient has had a tumor surgically removed and molecularly characterized through sequencing, we possess a precise catalog of that tumor's somatic mutations. From that point onward, a blood sample can be used to analyze circulating cell-free DNA (cfDNA), which contains fragments released into the bloodstream both from normal cells undergoing routine turnover and, crucially, from tumor cells. Tumor cells—especially those undergoing rapid proliferation, apoptosis, or necrosis—shed DNA into the circulation at higher rates than healthy tissues. When a metastatic focus begins to grow, even if it is not yet detectable by imaging, dying tumor cells release fragments of their mutated DNA, creating a small but measurable signal within the background of the patient's own cfDNA. By sequencing this mixture and filtering the variants against the known mutational profile of the original tumor, clinicians can identify the reappearance of tumor-derived DNA, providing an early indication of recurrence or metastasis.

Several studies have explored this strategy and observed that the technique could detect circulating tumor DNA (ctDNA) just a few weeks after initial treatment. Although these studies had a small sample size, the results indicated a clear enrichment: patients who later experienced recurrences were those in whom ctDNA was detected early. When comparing detection times across broader studies, the advantage is significant. For example, ctDNA was detected several months or even years before conventional monitoring techniques could identify any abnormalities. In one specific study, at 30 months, none of the traditional methods had identified recurrences, while ctDNA analysis had successfully detected them in about half of the cases. This early detection enabled the timely commencement of treatment, thereby significantly boosting disease-free survival rates in patients who were monitored using ctDNA.

In conclusion, the ongoing advancements in cfDNA technologies represent a paradigm shift in cancer management. As the precision of these molecular tools continues to improve, the potential to revolutionize early detection, and proactive treatment strategies will undoubtedly save lives, providing hope that the battle against cancer can be fought more effectively than ever before.

Oncogenomics Unites Clinical Genomics

The study of cancer genomics represents, in many ways, the point where all the concepts developed throughout this book converge—and simultaneously one of the areas in which clinical genomics is having, and will continue to have, its most transformative impact. The foundations of clinical genomics (Chap. 1), the structure and function of the human genome (Chap. 2), the technologies that allow us to read DNA (Chap. 3), and the essential tools of bioinformatics (Chap. 4) are all prerequisites for understanding the molecular landscape of tumors, and as seen in this chapter they display several peculiarities when applied to the tumor genome. Likewise, the principles behind variant discovery and interpretation (Chap. 5), their application to Mendelian disease diagnosis (Chap. 6), the formal ACMG classification framework (Chap. 7), and the translation of genomic findings into clinical reports (Chap. 8) reappear in oncology, adapted to the unique behavior of somatic variants. Even the concepts from population genomics (Chap. 9), that model disease susceptibility and allow evaluating the risk of future disease development, and the ethical, legal, and social considerations of genomic medicine (Chap. 10), become particularly salient in a field where genetic information can affect not only patients but also their families and broader health policies. In this sense, cancer genomics synthesizes all prior knowledge, deploying these tools with unprecedented clinical power. It stands today as one of the most rapidly expanding and promising frontiers of clinical genomics, driven by technological innovation, increasingly accurate risk models, and the development of targeted therapies rooted in the molecular understanding of each tumor. Ultimately, oncogenomics not only brings together the journey covered in previous chapters, but it also clearly illuminates the path ahead for the future of precision medicine.

References and Suggested Further Readings

Hanahan D. Hallmarks of cancer: the next generation, Cell. 2011;144(5):646–74, https://doi.org/10.1016/j.cell.2011.02.013

Chakravarty D, Solit DB. Clinical cancer genomic profiling. Nat Rev Genet. 2021;22(8):483–501. https://doi.org/10.1038/s41576-021-00338-8.

Li MM, Standards and guidelines for the interpretation and reporting of sequence variants in cancer: a joint consensus recommendation of the Association for Molecular Pathology, American Society of Clinical Oncology, and College of American Pathologists, J Mol. Diagn. 2016;19(1):4–23 https://doi.org/10.1016/j.jmoldx.2016.10.002

Rose M, Burgess JT, O'Byrne K, Richard DJ, Bolderson E. PARP inhibitors: clinical relevance, mechanisms of action and tumor resistance. Front Cell Dev Biol. 2020;8:564601 https://doi.org/10.3389/fcell.2020.564601.

Tyrer J, Duffy SW, Cuzick J. A breast cancer prediction model incorporating familial and personal risk factors. Stat Med. 2004;23(8):1111–30. https://doi.org/10.1002/sim.1668.

Lee A, Mavaddat N, Cunningham A, Carver T, et al BOADICEA: a comprehensive breast cancer risk prediction model incorporating genetic and nongenetic risk factors. Genet Med. 2019;21(8):1708–18. https://doi.org/10.1038/s41436-018-0406-9.

FSC
www.fsc.org
MIX
Papier | Fördert gute Waldnutzung
FSC® C083411

Kolibri 360 GmbH
Ferdinand-Jühlke-Straße 7
99095 Erfurt, Deutschland
produktsicherheit@kolibri360.de